John A. A. Sillince · Maria Sillince

Molecular Databases for Protein Sequences and Structure Studies

An Introduction

With 27 Figures

Springer-Verlag
Berlin Heidelberg NewYork
London Paris Tokyo
Hong Kong Barcelona Budapest

Dr. John A. A. Sillince
Lecturer in Management Information Systems
British Sheffield University
Management School
Crookesmoor Building
Conduit Road
Sheffield S10
1FL, UK

Dr. Maria Sillince
Assistant Subject Librarian
Wolverhampton Polytechnic
Wolverhampton
UK

ISBN-13: 978-3-642-76811-8 e-ISBN-13: 978-3-642-76809-5
DOI: 10.1007/978-3-642-76809-5

Library of Congress Cataloging-in-Publication Data
Sillince, John
Molecular Databases for protein sequences and structure studies :
an introduction / J. A. A. Sillince, M. Sillince
Includes bibliographical references and index.

1. Amino acid sequence--data processing.
2. Proteins--Analysis--Data processing.
3. Expert systems (Computer science)
I. Sillince, M. (Maria)
II. Title.
QP551.S56 1991
574.19'245'0285--dc20 91-25898

Softcover reprint of the hardcover 1st edition 1991

51/3020-5 4 3 2 1 0 Printed on acid-free paper

Acknowledgements

We are very grateful to all those scientists who replied to the questionnaire discussed in the last chapter of this book. Although lack of space prevents listing them all, the replies of the following were particularly helpful: Professor T.R. Blundell, Dr. A. Sali, Dr. J.P.Overington, Dr. M.S. Johnson, Dr. J.M. Thornton and Dr. S.P. Gardner at the Department of Crystallography at Birkbeck College, London University; Dr. D.A. Clark, Dr. G.J. Barton and Dr. C.J. Rawlings at the Biomedical Computing Unit, Imperial Cancer research Fund, London; Dr. E.M. Mitchell , Dr.P.J. Artymiuk, Dr.D.W. Rice, and Dr. P. Willet at the Department of Information Studies, University of Sheffield; Dr.S.Karlin at the University of California for making available copies of papers prior to publication; Dr. P.E. Jansson, Dr. L. Kenne, and Dr. G. Widmalm of the Department of Organic Chemistry, University of Stockholm.

Similarly for the latest information on biomolecular databanks we would like to thank Dr. R.F. Doolittle, Dr. W.C. Barker and Dr. D.F. Feng at the Centre for Molecular Genetics, University of California, San Diego; Dana Smith at the Complex Carbohydrate Research Center, University of Georgia; Dr. J.R.Lawton, Dr. F.A. Martinez and Dr. C. Burks at Los Alamos; also information scientists at Fisons Pharmaceuticals, Loughborough, and at Courtaulds, Coventry; also we are grateful to Professor J. Meadows and Dr. J.F.B. Rowland at the Department of Library and Information Studies at Loughborough University for their assistance in planning and criticising parts of the manuscript; also we would like to thank librarians at Loughborough and Warwick University Libraries and at Coventry and Wolverhampton Polytechnic Libraries; and at the British Library Document Supply Centre, Weatherby.

Also we would like to acknowledge the kind permission of the following for allowing us to reproduce material: Figure 1.1 is reprinted with permission by Watson J.D., "Molecular biology of the gene", Copyright (c) 1976, Benjamin/ Cummings Publishing Company; Figure 1.2a is reprinted with permission from McGammon J.A., and Harvey S.C.,Copyright (c) 1987, "Dynamics of proteins and nucleic acids", Cambridge University Press; Figure 1.2b is reprinted with permission from McGammon J.A., and Harvey S.C., Copyright (c) 1987, "Dynamics of proteins and nucleic acids", Cambridge University Press; Figure 1.3 is reprinted with permission from McGammon J.A., and Harvey S.C.,Copyright (c) 1987, "Dynamics of proteins and nucleic acids", Cambridge University Press; Figure 1.4 is reprinted with permission from F.H.C.Crick, "The genetic code", in "Recombinant DNA", Copyright (c) 1978, W.H.Freeman, Figure 1.5. is reprinted with permission from Bishop M.J., Ginsburg M., Rawlings C.J., and Wakeford R., "Molecular sequence databases", in Bishop M.J., and Rawlings C.J., (Eds) "Nucleic acid and protein analysis: a practical approach", Copyright (c) 1987, IRL Press, by permission of Oxford University Press; Figure 1.6 is reprinted with permission from Brown D.D., "The isolation of genes", in "Recombinant DNA", Copyright (c) 1978, W.H.Freeman, Figure 1.7 is reprinted with permission from Bishop M.J., Ginsburg M., Rawlings C.J., and Wakeford R., "Molecular sequence databases", in Bishop M.J., and Rawlings C.J., (Eds) "Nucleic acid and protein analysis: a practical approach", Copyright (c) 1987, IRL Press,by permission of Oxford University Press; Figure 2.1 is reprinted with permission from Nucleic Acids Research, Vol. 11, Hawley D.K., and

McClure W.R., Copyright (c) 1983, IRL Press, by permission of Oxford University Press; Figure 4.4 is reprinted with permission from Ash J.E., Chubb P.A., Ward S.E., Welford S.M., and Willett P., "Communications storage and retrieval of chemical information", Copyright (c) 1985, Ellis Horwood; Figure 4.5 is reprinted with permission from Ash J.E., Chubb P.A., Ward S.E., Welford S.M., and Willett P., "Communications storage and retrieval of chemical information", Copyright (c) 1985, Ellis Horwood; Figure 4.6 is reprinted with permission from Lynch M.F., Harrison J.M., Town W.G., and Ash J.E., "Computer handling of chemical structure information", Copyright (c) 1971, Macdonald; Figure 4.7 is reprinted with permission from Computers and Chemistry, Vol.13, Haines R.C., "Computer graphics for processing chemical substance information", Copyright (c) 1989, Pergamon Press PLC; Figure 5.1 is reprinted with permission from Database, Vol. 10, Barnard J.M., "Online graphical searching of Markush structures in patents", Copyright (c) 1987, Online Inc; Figure 5.2 is reprinted with permission from Lynch M.F., Harrison J.M., Town W.G., and Ash J.E., "Computer handling of chemical structure information", Copyright (c) 1971, Macdonald; Figure 7.1 is reprinted with permission from Keil B., "Cooperation between databases and the scientific community", in Doolittle R.F., (Ed) "Molecular evolution: computer analysis of protein and nucleic acid sequences", Copyright (c) 1990, Methods in Enzymology, Vol. 183, Academic Press; Figure 7.3 is reprinted with permission from Bishop M.J., Ginsburg M., Rawlings C.J., and Wakeford R., "Molecular sequence databases", in Bishop M.J. and Rawlings C.J. (Eds) "Nucleic acid and protein analysis: a practical approach", Copyright (c) 1987, IRL Press, by permission of Oxford University Press; Figure 8.1 is reprinted with permission from Database, Vol. 121, Allen F.C., and Ferrell W.R., "Numeric databases in science and technology: an overview", Copyright (c) 1989, Online Inc; Figure 8.2

is reprinted by permission from Nature Vol. 335, pp. 745-748, Copyright (c) 1988, MacMillan Magazines Ltd; Figure 9.1 is reprinted with permission from Doolittle R.F., "Of ORFs and URFs: a primer on how to analyse derived amino acid sequences", Copyright (c) 1986, University Science Books; Figure 9.2 is reprinted by permission from Nature Vol. 296, pp. 171-173, Copyright (c) 1982, MacMillan Magazines Ltd; Figure 9.3 is reprinted with permission from Doolittle R.F., "Of ORFs and URFs: a primer on how to analyse derived amino acid sequences", Copyright (c) 1986, University Science Books; Figure 10.2 is reprinted with permission from Thornton J.M. and Taylor W.R., "Structure prediction", in Findlay J.B.C. and Geisow M.J., (Eds), "Protein sequencing: a practical approach", Copyright (c) 1989, IRL Press, by permission of Oxford University Press.

Preface

Molecular information is now too vast to be acquired without the use of computer-based systems, which can select according to supplied criteria. Developments in programming have created the ability to extend molecular science in ways that would have been impossible without their help. New databases are being established which enable previously unanswerable questions to be considered. One of these questions is whether or not one can predict the three dimensional structure of a protein from information about its sequence of amino acids.

In order to help the reader to understand this new field, several topics are explained in this volume. The structure and function of proteins and nucleic acids are described, in order to emphasise the way in which three dimensional structure reflects a protein's role in the organism. Also it is important to consider what is involved in molecular data, and how it is represented and registered in software and on the screen. Another aspect to consider is how computer-based research tools are used in molecular science, in particular for manipulating sequence and structure information. Sequence and structure are at the centre of research problems in molecular science, in the identification of a new protein (14000 are known so far) or its three dimensional structure (only 400 are known so far), in patent writing and patent searching, and in modelling proteins.

There is also a description of the state of the art in what databanks exist, both for sequences and for structures, and what types of system are available for using them, for modelling, for searching, and for integrating the operations of the online database and the local system such as a PC in a laboratory. New developments in knowledge-based systems and database technology are described. The case study on protein structure prediction, which includes developments in the specification of an expert system for such a problem, is intended to exemplify the integrated nature of modelling and search (both computer-based) and laboratory experiment in molecular science.

Contents

List of Figures

Chapter 1:

Introduction

1.1. AIMS OF BOOK

Very little has been written about online molecular databases from the biologist's and the computer scientist's point of view. What little has been written is mainly intended for the research worker rather than for the scientist coming to online molecular databases for the first time. Moreover, although books do exist about chemical databases, little which has been written about molecular databases goes very deeply into the important chemical structures involved. There is thus a gap in the literature in terms of a book which (1) shows the broad scientific context of molecular databases, (2) provides an introduction to the computational aspects, (3) shows the latest developments in the field, and (4) shows the chemical background of molecular information.

This book aims to fill part of this gap by drawing attention to the main databanks in the field of biochemistry, molecular biology, biotechnology and genetics. The use of these databanks needs up-to-date subject knowledge on a narrow field and a wide general or multidisciplinary knowledge of adjacent fields. These adjacent fields will be discussed here. Also it is intended to give a few detailed examples of databank use in one of the most interesting and advanced problems of molecular biology: the three-dimensional structure and biological function of proteins. It tries to identify those information and research tools which are used to store, analyse and compare sequence data with atomic coordinates resulting from X-ray (or recently neutron diffraction) data, in order to derive biological information on a given protein.

The role of computer applications in data handling and analysis will be shown through examples of databank searches from the literature. Computer applications extend the investigator's knowledge and so can be considered as research as well as information retrieval tools. The main sequence and structure databanks will be described in some detail, focussing on recent developments.

Finally the book tries to assess the feasibility of an expert system for three-dimensional structure prediction of proteins. This is based upon a questionnaire sent to university research scientists engaged in protein structure studies.

1.2. THE STRUCTURE AND ROLE OF PROTEINS AND NUCLEIC ACIDS

It is necessary now to define the various levels at which protein structure is studied.

1.2.1. Protein structure

Proteins are large organic molecules (macromolecules) built of more than 100 peptide units formed from the 20 different amino acids (Figure 1.1).

Polypepides are smaller, and have less than 100 units. The average molecular weight (MW) of amino acids is between 110 and 120. The molecular weight of proteins or polypeptides depends on the number of units

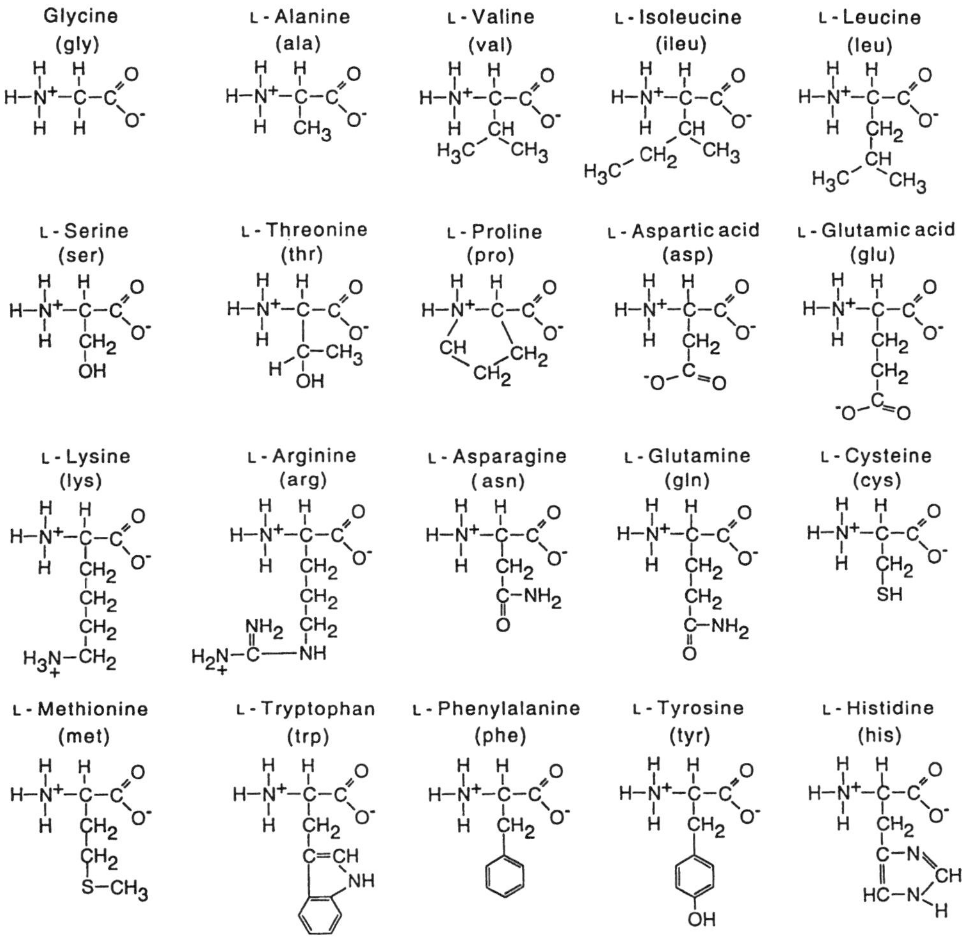

Figure 1.1. The 20 common amino acids found in proteins. (Watson, 1976, p.72).
Reprinted with permission by Watson J.D., "Molecular biology of the gene", Copyright (c) 1976, Benjamin/Cummings Publishing Company.

joined together, which is not more than 300-500 in a single chain. Protein molecules consist of one or a small number of polypeptide chains folded compactly into a characteristic three-dimensional structure. These are complemented in some cases by one or more prosthetic groups (e.g. metal ions or organic molecules).

Intensive study of their structure and function has made it necessary to define several levels of structure such as primary, secondary, super-secondary, tertiary and quaternary structure.

1.2.2. Primary structure

This is defined by the amino acid sequence of the polypeptide chain which arises from the formation of covalent bonds (called peptide bonds) between amino acids as in Figure 1.2a.

1.2.3. Secondary structure

Secondary structure refers to the local spatial arrangement of the backbone without regard to the conformation of the side chains. But sometimes it can mean the overall arrangement of the entire chain, including the side chain of the amino acids. This 'spatial arrangement' of the chain is described as the secondary and the tertiary structure, or as the conformation of proteins (Ghelis and Yon, 1982).

On the other hand there are many publications referring to regular secondary structures such as alpha-helix in which the peptide oxygen of residue i forms a hydrogen bond with the peptide nitrogen of residue i+4 as is shown in Figure 1.3a. Another structure, the beta-sheet, which has a similar function, is shown in Figure 1.3b. In the beta-sheet extended strands of the polypeptide chain (parallel or anti-parallel) are cross-linked by hydrogen bonds between their peptide groups.

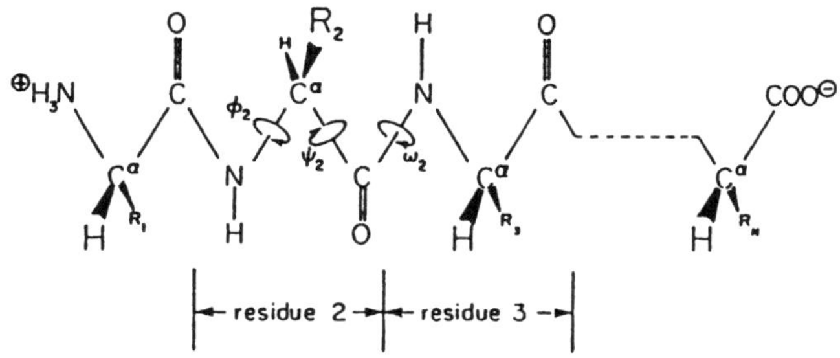

Figure 1.2a.

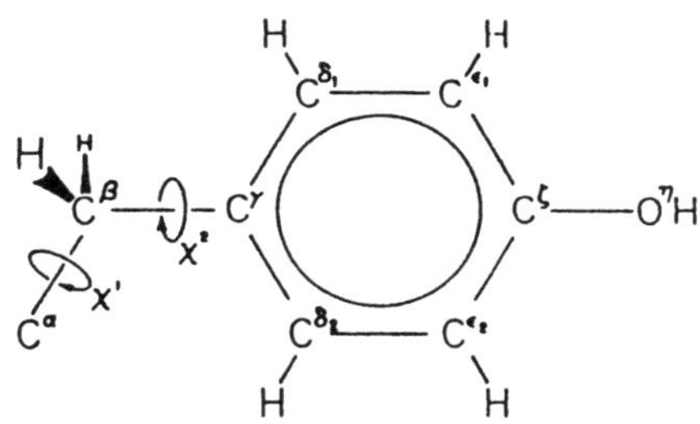

Figure 1.2b.

Figure 1.2. The structure of a polypeptide chain. The backbone or main chain is shown in Figure 1.2a, where the covalent bonds and bond angles are rather rigid, but sizeable rotations can occur around certain bonds. The dihedral angles ϕ_i and ψ_i measure the torsion about the rotationally permissive bonds in the backbone of residue i. The dihedral angles ω_i exhibit little variation because the C-N bond has partial double bond character; each peptide group (CO-NH) and its adjoining C^{α} atoms therefore tend to remain in a common plane. In Figure 1.2b, the Ri represent the sidechains, one of which, a tyrosine sidechain, is shown.
Reprinted with permission from McGammon J.A., and Harvey S.C., Copyright (c) 1987, "Dynamics of proteins and nucleic acids", p.12, Cambridge University Press.

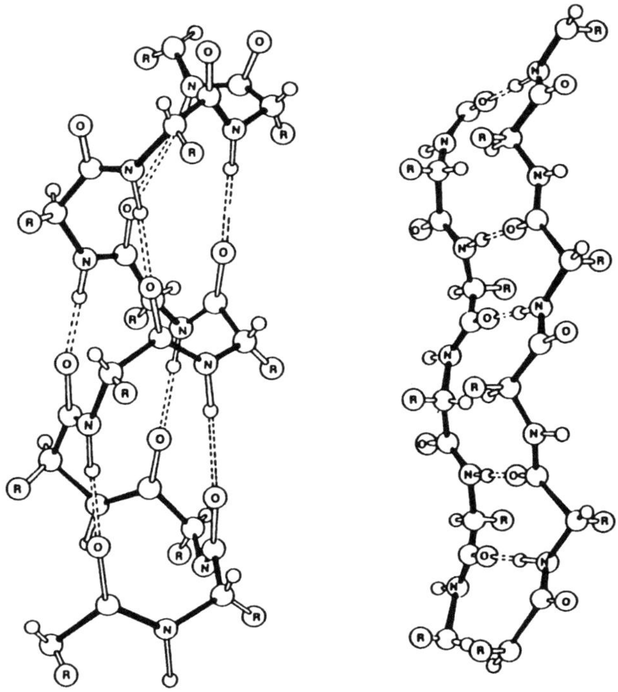

(a) (b)

Figure 1.3. The alpha-helix (a)
and the beta-sheet (b).
(McCammon and Harvey, 1987, p.15).
Reprinted with permission from McGammon J.A.,
and Harvey S.C.,Copyright (c) 1987, "Dynamics of proteins
and nucleic acids", Cambridge University Press.

1.2.4. Super-secondary structure

This means a group of secondary structure elements that "are generally sequentially adjacent and pack together in a regular way" (Taylor, 1987). These motifs dominate tertiary fields of most globular proteins and can be divided into five structural classes : all-alpha, all-beta, alternating alpha/beta, alpha+beta, and random (see p.155 in Thornton and Taylor, 1989). Typical super-secondary structures identified so far include the beta-alpha-beta unit, the beta-hairpin, the beta-meander, and the four-helical bundle.

1.2.5. Tertiary structure

The usual meaning of tertiary structure is the whole three-dimensional structure, i.e. the overall conformation of the protein molecule (side chain conformation). The side chains of the amino acid residues which can be either soluble in water (hydrophilic) or less soluble (hydrophobic), play an important role in the folding of the polypeptide chain which not only contributes to its native three-dimensional structure but to its previously-mentioned biological function.

1.2.6. Quaternary structure

This is defined by the geometry of subunit assembly. Many large molecules have two or more chains or subunits. For example, haemoglobin has four chains (two alpha + two beta) of which two are identical. The alpha chain has 141 amino acid residues while the beta chain has 146. Another example is the recently-reviewed ribulose-1,5-biphosphate carboxylase/oxygenase (Rubisco), the enzyme that is responsible for the incorporation of carbon dioxide into organic matter. Rubisco from higher plants and most algae is composed of 8 large and 8 small subunits (L8 S8) (Knaff, 1989; Schneider *et al.*, 1990).

1.2.7. DNA structure

The information that directs amino acid sequences in proteins is carried by the chain-like molecule of deoxyribonucleic acid (DNA) (see Figure 1.4). The cells of bacteria have often only a single chain, while the cells of higher organisms (plants, fruit fly, mammals) have many clustered together in chromosomes. Watson (1976, p.702) defines a gene as "a stretch along a chromosome that codes for a functional product either RNA or its translation product, a polypeptide".

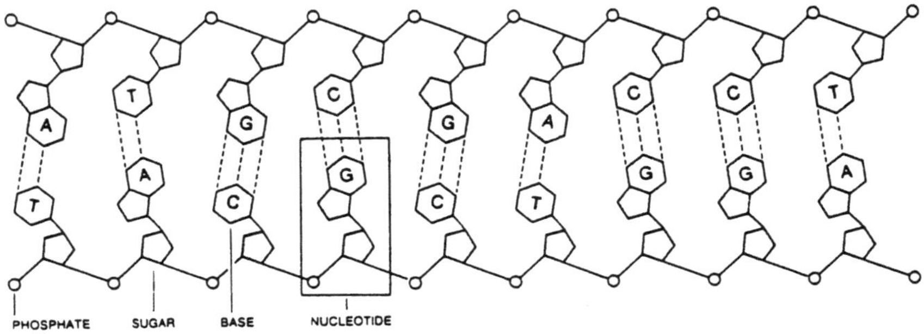

Figure 1.4. The DNA molecule, showing the chain of nucleotides, each designated by the initial letter of the organic base it contains. (Source: Brown, 1978, p.82).
Reprinted with permission from Brown D.D., "The isolation of genes", p.82 in "Recombinant DNA", Copyright (c) 1978, W.H.Freeman.

The DNA molecules are formed by the union of smaller parts called nucleotides (see Figure 1.5). The nucleotides of DNA contain deoxyribose (a five-carbon sugar), a phosphate group and one of the four bases. Two of them are purines, adenine (A) and guanine (G), and two are pyrimidines, thymine (T) and cytosine (C). The backbone of the molecule is formed by the sugar and phosphate group and is very regular, while the order of the bases along the chain varies and their sequence along the length of the DNA molecule (usually more than 100 bases) determines the structure of the particular protein for which the gene codes (see Figure 1.4). One of the most important features of DNA is that it usually consists of two complementary chains or strands twisted around each other in the form of a double helix (Watson, 1976, p.208). The two chains are joined together by hydrogen bonds between pairs: thymine is always paired with adenine and cytosine with guanine.

The hydrogen bonding between pairs is referred to as secondary structure and the arrangement of the double helices (the right-handed

Code	Nucleotides
A	adenine
C	cytosine
G	guanine
T	thymine
U	uracil
R	purine (A or G)
Y	pyrimidine (C or T/U)
M	A or C
W	A or T/U
S	C or G
K	G or T/U
D	A,G, or T/U
H	A,C, or T/U
V	A,C, or G
B	C,G, or T/U
N	A,C,G, or T/U

Figure 1.5. Codes for nucleotides, including uncertainties.
(Source: Bishop et al., 1987, p.88).
Reprinted with permission from Bishop M.J., Ginsburg M., Rawlings C.J., and Wakeford R., "Molecular sequence databases", in Bishop M.J., and Rawlings C.J., (Eds) "Nucleic acid and protein analysis: a practical approach", Copyright (c) 1987, IRL Press.
By permission of Oxford University Press.

A-DNA and B-DNA and the left-handed Z-DNA) indicates local structural conformation (McCammon and Harvey, 1987).

The genetic code is often referred to as a 'dictionary' used by the cell to translate from the four-letter language of nucleic acids to the 20-letter language of proteins. The cell can translate in one direction only, from nucleic acid to protein but not from protein to nucleic acid. During this process of protein synthesis the message contained in DNA is first transcribed into the similar molecule called 'messenger' ribonucleic acid (mRNA). mRNA too has four bases as side groups: three are the same as in DNA (A,C,G) but the fourth is uracil (U) instead of thymine (T). The other difference is in the structure of the sugar component, which is ribose instead of deoxyribose. In this first transcription of the genetic message the code letters A,C,G,T of DNA become respectively U,G,C,A of mRNA, because the complementarity determines the sequence of the mRNA. The amino acids are transported to the ribosome part of the cell where the synthesis takes place by a particular form of "transfer" RNA (tRNA) which carries an anticodon that forms a temporary bond with one of the codons in messenger RNA (see Figure 1.6). (A codon is a triplet of bases representing an amino acid.)

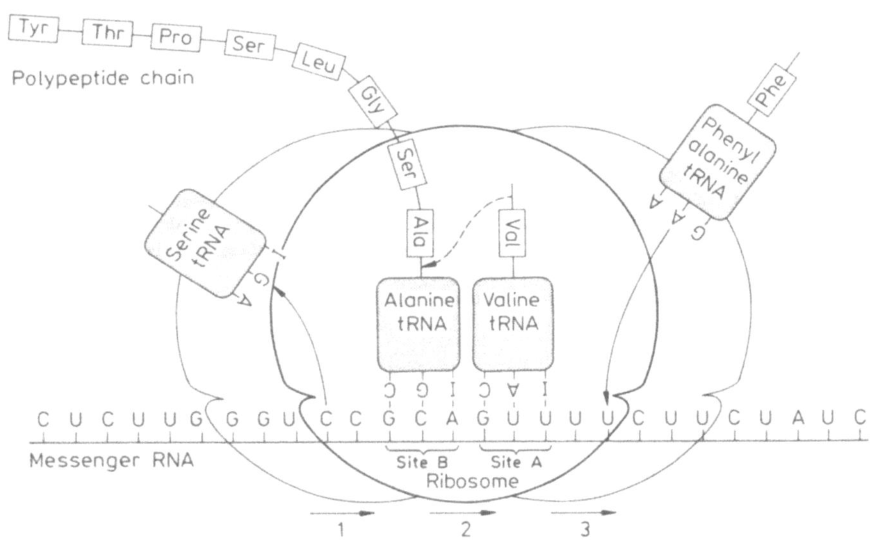

Figure 1.6. Synthesis of protein molecules. (Source: Crick, 1978a, p.21). Reprinted with permission from Crick F.H.C., "The genetic code", p.21 in "Recombinant DNA", Copyright (c) 1978, W.H. Freeman.

1.3. THE NATURE OF MOLECULAR DATA AND ITS REPRESENTATION

At this point it is necessary to describe the various representations of sequence and structure data.

1.3.1. Sequence data and its representation

The 20 amino acids are each represented in machine-readable form as a single letter, while their representation in three-letter form is mnemonically more pleasing to human users (see Figure 1.7). So for example alanine is Ala or A, arginine is Arg or R, asparagine is Asn or N and so on.

Codes		Amino Acid
A	Ala	alanine
B	Asx	aspartic acid or asparagine
C	Cys	cysteine
D	Asp	aspartic acid
E	Glu	glutamic acid
F	Phe	phenylalanine
G	Gly	glycine
H	His	histidine
I	Ile	isoleucine
K	Lys	lysine
L	Leu	leucine
M	Met	methionine
N	Asn	asparagine
P	Pro	proline
Q	Gln	glutamine
R	Arg	arginine
S	Ser	serine
T	Thr	threonine
V	Val	valine
W	Trp	tryptophan
X	X	any amino acid
Y	Tyr	tyrosine
Z	Glx	glutamine or glutamic acid

Note: These codes conform to the IUPAC-IUB standard.

Figure 1.7. The one-letter and three-letter codes for amino acids. (Source : Bishop et al., 1987, p.90.)
Reprinted with permission from Bishop M.J., Ginsburg M., Rawlings C.J., and Wakeford R., "Molecular sequence databases", in Bishop M.J., and Rawlings C.J., (Eds) "Nucleic acid and protein analysis: a practical approach",

Sequence data describes the primary structure of a given protein polypeptide, i.e. the order in which amino acid residues occur. In spite of new techniques in protein purification and sequencing such as reversed phase HPLC (high performance liquid chromatography) and gel electrophoresis, amino acid sequencing is much slower than nucleic acid sequencing. So the main body (80 - 90%) of protein sequence data comes from nucleic acid sequencing using the genetic code. This process is called "sequence translation". It means "the formation of a protein sequence by replacing each triplet codon of the nucleic acid sequence with its corresponding amino acid residue" (Stockwell, 1987). As mentioned above, in DNA the base nucleotides are either guamine (G), thymine (T), adenine (A), and cytosine (C), while in messenger RNA thymine (T) is replaced by uracil (U). Although sequences of mitochondrial origin require different codes, in this context is meant the "standard genetic code" which most DNA sequences require. Figure 1.8 gives the genetic code. Using it one can translate from a three-letter codon into an amino acid. For example TTT gives the amino acid Phe (phenylalanine) and TCT gives Ser (serine). Note that there are 20 amino acids (21 when one includes the chain terminator) and 64 different permutations from the 3 letters of the codon . Thus in some cases the same amino acid can be coded by different codons. Figure 1.8 refers to DNA: for RNA translation one must replace thymine (T) by uracil (U). This list of 5 bases (4 for DNA, 4 for RNA) can be extended to include uncertanties (for example M stands for A or C) to a list of 16 base codes for nucleotides (see Figure 1.5) (Bishop et al, 1987, p.88).

First Position	Second Position				Third Position
	T	C	A	G	
T	Phe	Ser	Tyr	Cys	T
	Phe	Ser	Tyr	Cys	C
	Leu	Ser	Term	Term	A
	Leu	Ser	Term	Trp	G
C	Leu	Pro	His	Arg	T
	Leu	Pro	His	Arg	C
	Leu	Pro	Glu	Arg	A
	Leu	Pro	Glu	Arg	G
A	Ileu	Thr	Asn	Ser	T
	Ileu	Thr	Asn	Ser	C
	Ileu	Thr	Lys	Arg	A
	Met	Thr	Lys	Arg	G
G	Val	Ala	Asp	Gly	T
	Val	Ala	Asp	Gly	C
	Val	Ala	Glu	Gly	A
	Val	Ala	Glu	Gly	G

Figure 1.8. The genetic code.

1.3.2. Structure data and its representation

The secondary structure of a protein as it was defined above describes the spatial arrangement of its main-chain atoms. This information is mainly available from X-ray and neutron diffraction studies although the structure of a few oligopeptides have been determined by nuclear magnetic resonance (NMR) techniques in solution recently.

To grow a large protein crystal for X-ray treatment is the first and most time-consuming part of the process which may take several months. Data is mainly in the form of x-y-z atomic coordinates, structure factors and angles. Data collection starts after the characterisation of the diffraction pattern which gives information about the resolution of the electron density and the size and symmetry of the unit cell of the crystal. The diffraction pattern also determines the method of data collection (Eisenberg and Hill, 1989, Table 1). The use of synchrotron X-ray sources and area detectors which are used for collecting intensities of the diffraction spots of larger unit cells are improvements which have speeded up X-ray data collection. This has increased the quantity of available data recently to about 400 sets of coordinates, about 100 of which were deposited in 1988. Phase determination (using heavy atoms that bound protein crystals) and calculation of electron density are the next steps which are followed by building a model of the protein by means of the electron density using computer graphics. The final step is improving the model by atomic refinenment and interpreting the final model. Refinement methods are becoming more powerful with the introduction of restrained least squares methods and combined X-ray and molecular-dynamics techniques. The strengths of these refinement methods (which cover processes which occur in under a nanosecond - or a billionth of a second) is that very detailed models can be built, where every atom and its potential functions can represent all forms of inter-atomic interactions. At the same time this very detailed representation produces a serious limitation. For example a 100 pico second (a pico second is ten to the minus twelve of a second) molecular dynamics simulation of a macromolecule of 30,000 molecular weight is computationally very expensive to solve. This means that important biological problems involving large molecules and motions on time scales longer than one nanosecond cannot yet be treated easilly (McCammon and Harvey, 1987,p.157).

1.4. THE IMPORTANCE OF PROTEIN STRUCTURE AND FUNCTION STUDIES

Proteins and polypeptides have always been at the centre of interest because they are some of the main components of the living cell. Their importance is partly due to the diversity of their roles in the functioning of the organism. One group of proteins function as elements in the muscle (actin, myosin), others catalyse biochemical reactions (lysozyme, trypsin), others are transport proteins (cytochrome c, ferredoxins, haemoglobins) or antibodies (immunoglobulins), and there are many more.

Due to developments in purification and molecular weight determination methods, more and more proteins have been studied. Their primary structure (the sequence of the amino acids) revealed that proteins with the same function in different organisms are very similar. For example the position of cysteine residues which bind the iron atoms are identical in the ferredoxins which come from green sulphur bacteria, fermentative anaerobic bacteria, and purple bacteria. Similarly the active centres of ferredoxins which come from blue-green algae, green algae, and higher plants all contain a single iron-sulphur cluster [2Fe-2S] which participates in electron transfer during biological oxido-reduction. Not only ferredoxins but also c-type cytochromes from different organisms (such as horse, tuna and some photosynthetic and aerobic bacteria) show that the haem-containing active site occupies the same position in all the cytochrome-c molecules (Rao et al , 1981).

These data and many similar studies have shown that a number of proteins can be used to study evolutionary relatedness. It had been thought that an early ancestor had a relatively small genome coding for a relatively narrow range of prototypic proteins and the present proteins are the results of past gene duplications and subsequent divergence resulting from gradual amino acid replacement (Doolittle, 1989).

However, there followed a methodological breakthrough in nucleic acid sequencing. This, together with further technical developments not only increased the quantity of data enormously but improved the possibility of a better understanding of the relationship between structure and function of different proteins. It has been found that the biological activity of a protein depends on its three-dimensional structure. The correlation of sequence data with information obtained from genetic studies in identifying inherited diseases such as sickle cell anaemia provides another example of the importance of protein structure studies. Sickle cell anaemia is due to a single base change in one of the codons of the beta chain of the haemoglobin, where each codon specifies a particular amino acid in the protein product. Instead of GTG, the codon for valine, the triplet changed to GAG which codes for glutamic acid. This mutation changed the structure of haemoglobin (it forms long rod-like filaments that changes the shape of the red blood cells). The haemoglobin's function was modified by this, reducing the amount of oxygen which the haemoglobin could carry. Similarly, structure studies of viral proteins and tumour necrosis factors can help us to understand the functional relationship of different proteins (Jones et al , 1989) and have the potential to offer treatment. This field is taking over from evolution studies although the two still overlap and complement one another.

The examples above and recent studies on the role of carbohydrate (sugar) groups in biological recognition emphasise the central place of chemical information in modern molecular biology. However, the example below also exemplifies the crucial role which computer science is also beginning to assume in molecular science.

Perhaps the most ambitious molecular science project is the Human Genome Project, which aims to sequence the human genome. It is of such vast proportions that the research effort and available databases will have to be shared on an international scale. It has been estimated that the project will eventually cost $10 billion since there are 3 billion base pairs in the human genome and each base pair currently costs between $3 and $5 to sequence. A special agency (the Human Genome Organisation or HUGO) has been established to coordinate the international research effort. Although the U.S. Government was the prime mover in the project, large funds have recently been made available by Japan also. The Human Genome Project will increase research even faster not only in human genetics and biochemistry but in plants and animals too. This field will attract not only many more molecular biologists and biochemists as von Heijne (1987) predicts but also information specialists and computer programmers who will join them and who will have to find new methods of data storage and retrieval to deal with the need for very large databases. Unless advances in database technology are made the information discovered by the project will not be useable. The problem is not so much the secondary storage space (CDs of 500 million bytes now exist and with each base occupying a byte of memory the whole genome could fit on 6 or 7 disks) but rather speeding up search methods.

1.5. REFERENCES

Bishop M.J., Ginsburg M., Rawlings C.J., and Wakeford R., 1987, Molecular sequence databases, 83-114 in Bishop M.J., and Rawlings C.J., (Eds), Nucleic acid and protein sequence analysis: a practical approach, I.R.L. Press, Oxford.

Brown D.D., 1978, The isolation of genes 81-90 in Recombinant DNA, W.H.Freeman, San Francisco.

Crick F.H.C., 1978a, The genetic code III, 21-27 in Recombinant DNA, W.H. Freeman, San Francisco.

Crick F.H.C., 1978b, Nucleic acids, 6-11 in Recombinant DNA, W.H.Freeman, San Francisco.

Doolittle R.F., 1989, "Similar amino acid sequences revisited", Trends Biochem. Sci., 14, 244-245.

Ghelis C., and Yon J., 1982, Protein folding, Academic Press, New York.

Eisenberg D., and Hill C.P., 1989, "Protein crystallography: more surprises ahead", Trends Biochem. Sci., 14, 260-264.

Jones E.Y., Stuart D.I., and Walker N.P.C., 1989, "Structure of tumour necrosis factor", Nature, 338, 225-228.

Knaff D.B., 1989, "Structure and regulation of ribulose 1,5-bisphospate carboxylase," Trends Biochem. Sci., 14, 159-160.

McCammon J.A., and Harvey S.C., 1987, Dynamics of proteins and nucleic acids, Cambridge University Press, Cambridge.

Rao K.K., Hall D.O., and Cammack R., 1981, The photosynthetic apparatus, 150-202 in Gutfreund H., (Ed.), Biochemical evolution Cambridge University Press, Cambridge.

Schneider G., Knight S., Andersson I., Branden C.-I., Lindqvist Y., and Lundqvist T., 1990, "Comparison of the crystal structures of L2 and L8S8 Rubisco suggests a functional role for the small subunit", The EMBO Journal, 9(7), 2045-2050.

Stockwell P.A., 1987, DNA sequence analysis software, 19-45 in Bishop M.J., and Rawlings C.J., Nucleic acid and protein sequence analysis: a practical approach, I.R.L Press, Oxford.

Taylor W.R., 1987, Protein structure prediction , 285-322 in Bishop M.J., and Rawlings C.J., (Eds.), Nucleic acid and sequence analysis : a practical approach, I.R.L Press, Oxford.

Thornton J.M., and Taylor W.R., 1989, Structure prediction, 147-190 in Findlay J.B.C., and Geisow M.J., (Eds.), Protein sequencing: a practical approach, I.R.L. Press, Oxford.

von Heijne G., 1987, Sequences: where theoretical and experimental molecular biology meet, 151-152 in Sequence analysis in molecular biology, Academic Press, New York.

Watson J.D., 1976, Molecular biology of the gene, W.A.Benjamin, Menlo Park, California.

Chapter 2:

Computer-Based Research Tools for Molecular Science

2.1. THE USE OF COMPUTERS AND ONLINE FACILITIES IN SEQUENCING

Although there are many examples of computer applications in molecular science from data collection to modelling, we will only mention in detail those which relate to sequences. Even these applications would amount to several chapters because scientists have become more interested in further analysis of sequence data than before . Also new applications are being developed very rapidly.

2.1.1. Computer-based sequencing projects

Computer applications in the laboratory started with collection and management of primary data (starting with the order and location of the four nucleotides adenine, thymine, guanine and cytosine) from sequencing gels during large sequencing projects involving over 30,000 bases such as the bacteriophages lambda and T7 (Gingeras, 1983). The primary data is obtained from the gel by moving a light pen connected to a digitising device on a series of bands representing the nucleotides separated by their mobility, which depends on their molecular weight (autoradiograph). These projects usually involve pre-sequencing preparations such as the construction of restriction maps. Restriction maps are formed by fitting parts of the DNA segment after it has been cut with (i.e. digested by) restriction enzymes that can recognise particular bases (so one end of the fragment is known). From this a linear or circular map is formed.

Mapping programs generate all possible continuations of fragments which have been gained after several digestion steps and after grouping them by length. For example one program (Pearson, 1982) first considers all possible maps for the two single digests with the smallest number of fragments. All possible double digests from these maps are then evaluated against the data, a measure of the goodness of fit is calculated for each, and only those that have a calculated fit above some given value are retained. This process is repeated with possible maps from the third single digest, and so on, until at the end only one or a few maps remain (von Heijne, 1987).

The next step is to plan or work out the best way of conducting the sequencing. This strategy can be obtained using programs that have been developed to minimise the number of gels that have to be run and maximise the number of new bases read from each gel, given the maximum number of bases that the laboratory can read from a gel (Bach et al., 1982). Many of the integrated packages for DNA data management include digitising devices for direct entry of data about nucleotides from the gel to the computer. The data processing and other operations necessary in a sequencing project are described in detail by Staden (1987) and will not be covered here.

2.1.2. Computer-based sequence analysis

The first types of computer programs were written mainly to help organise and store new sequence data. Even these early applications included algorithms for detecting relatedness between two sequences such as the program by Needleman and Wunch (1970) which used matrix methods. Similarly programs by Sellers (1974, 1979) and Waterman et al, (1976) were some of the first ones used to calculate the evolutionary distances between sequences of different evolutionary origins.

With the sudden and exponential growth of sequence data it became apparent that sequence data contain a lot of information which is not immediately obvious. On the other hand molecular scientists can ask more questions related to sequence and other information due not only to the larger amount of data but also the number of longer sequences available. The development of a computer-recognisable language (DNA*) for molecular biologists by Friedland et al., (1982) in a meta-language called GENESIS was a further step (Gingeras, 1983).

Examples of sequence analysis include the identification and location of restriction sites, promoter sites, protein coding regions, (among many others) in large nucleotide sequences and the comparison of protein sequences stored in one or more databases already in order to determine similarity and homology. Identification is done by means of either of two methods of search: by signal or by content. Both methods build on knowledge gained from experiments which are analysed statistically.

Search by signal

Search by signal involves searching for similarities to a consensus sequence. Promoter sites are short conserved stretches of nucleotides involved in transcription. Promoter sequences are short specific sequencemotifs that regulate transcription by modulating the initial recognition and binding of RNA polymerase to DNA. For example, the consensus sequence for *Escherichia coli* promoters in the -10 region is the subsequence TATAAT (Hawley and McClure, 1983). This is shown in Figure 2.1, where each six-base subsequence is tested for consensus and where the

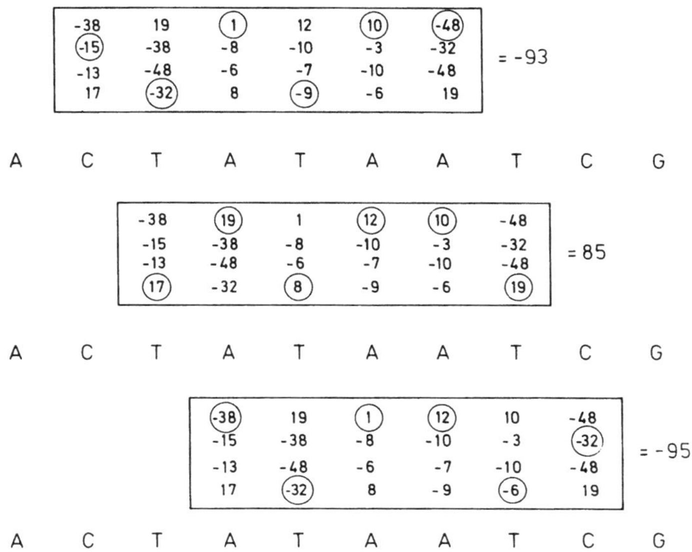

Figure 2.1. Matrix evaluation of a sequence.
(Source : Hawley and McClure, 1983).
Reprinted with permission from Nucleic Acids Research, Vol. 11, Hawley D.K., and McClure W.R., Copyright (c) 1983, IRL Press.
By permission of Oxford University Press.

highest consensus score calculated is 85. The four rows (top to bottom) are the bases A,C,G,T.

Each matrix contains an entry for each base and for each position in the sub-sequence. The evaluation of each potential site involves summing the elements that correspond to the sequence at that site. Figure 2.2(a) gives values of 60 to all -10 consensus sequences, 50 to all sites with one mismatch and so on down to zero for a site with no matches to the consensus. Figure 2.2(b) shows the effect of including more information about what is necessary to be a -10 promoter site. The consensus sequence is still evaluated to 60, but similar sequences whose mismatches are in different positions are evaluated differently. Mismatches in the most conserved positions provide the largest score reduction.

Position:	1	2	3	4	5	6
A	0	10	0	10	10	0
C	0	0	0	0	0	0
G	0	0	0	0	0	0
T	10	0	10	0	0	10

Figure 2.2(a) Matrix for -10 promoter region for consensus sequence. (Source: Stormo, 1987, p.233).

Position :	1	2	3	4	5	6
A	0	10	8	10	10	0
C	1	0	1	1	3	0
G	1	0	1	1	1	0
T	10	0	10	1	1	10

Figure 2.2(b) Matrix for -10 promoter region with different penalties for different mismatches to consensus. (Source: Stormo, 1987, p.233).

Position :	1	2	3	4	5	6
A	2	95	26	59	51	1
C	9	2	14	13	20	3
G	10	1	16	15	13	0
T	79	3	44	13	17	96

Figure 2.2(c) Matrix with elements proportional to the frequency of each base at each position in a collection of promoters.
(Source: Stormo, 1987, p.233).

Search by content

This method of finding coding regions is based on previous knowledge gained from comparing and evaluating sequences of known genes. For example, in a collection of E. coli genes 48% of the G's are in the first position, 20% are in the second position, and 32% are in the third position.The same can be done for every other base, the result being a standard base-position preference vector with 12 numbers.

Homologous protein sequences

In most protein matching exercises the user wishes to see where two protein sequences are similar, that is, at what relative positions in the amino acid sequences of the two proteins the similarity exists. Also he wishes to know how they are similar, in the sense of what functional attributes the two sequences have in common. And he must ascertain how significant the similarity is, relative to other similar sequences of the protein. This knowledge is gained by comparing every amino acid residue of the first protein with every amino acid residue of the second.

This involves a process of matching two protein sequences in terms of sets of characteristics which the two sequences share. These characteristics might be mutation substitution (Figure 2.3), hydrophobicity, propensity to form an alpha helix, beta sheet or beta turns, or fraction of sequence buried.

	A	R	N	D	C	Q	E	G	H	I	L	K	M	F	P	S	T	W	Y	V	B	Z	X
A	2	-2	0	0	-2	0	0	1	-1	-1	2	1	-1	-4	1	1	1	-6	-3	0	0	0	0
R	-2	6	0	-1	-4	1	-1	-3	2	-2	-3	3	0	-4	0	0	-1	2	-4	-2	-1	0	0
N	0	0	2	2	-4	1	1	0	2	-2	-3	1	-2	-4	-1	1	0	-4	-2	-2	2	1	0
D	0	-1	2	4	-5	2	3	1	1	-2	-4	0	-3	-6	-1	0	0	-7	-4	-2	3	3	0
C	-2	-4	-4	-5	12	-5	-5	-3	-3	-2	-6	-5	-5	-4	-3	0	-2	-8	0	-2	-4	-5	0
Q	0	1	1	2	-5	4	2	-1	3	-2	-2	1	-1	-5	0	-1	-1	-5	-4	-2	1	3	0
E	0	-1	1	3	-5	2	4	0	1	-2	-3	0	-2	-5	-1	0	0	-7	-4	-2	2	3	0
G	1	-3	0	1	-3	-1	0	5	-2	-1	-4	-2	-3	-5	-1	1	0	-7	-5	-1	0	-1	0
H	-1	2	2	1	-3	3	1	-2	6	-2	2	0	-2	-2	0	-1	-1	-3	0	-2	1	2	0
I	-1	-2	-2	-2	-2	-2	-2	-3	-2	5	2	-2	2	1	-2	-1	0	-5	-1	4	-2	-2	0
L	-2	-3	-3	-4	-6	-2	1	-4	-2	2	6	-3	4	2	-3	-3	-2	-2	-1	2	-3	-3	0
K	-1	3	1	0	-5	1	0	-2	0	-2	-3	5	0	-5	-1	0	0	-3	-4	-2	1	0	0
M	-1	0	-2	-3	-5	-1	-2	-3	-2	2	4	0	6	0	-2	-2	-1	-4	-2	2	-2	-2	0
F	-4	-4	-4	-6	-4	-5	-5	-5	-2	1	2	5	0	9	-5	-3	-3	0	7	1	-5	-5	0
P	1	0	-1	-1	-3	0	-1	-1	0	-2	-3	-1	-2	-5	6	1	0	-6	-5	1	-1	0	0
S	1	0	1	0	0	-1	0	1	-1	-1	-3	0	-2	-3	1	2	1	-2	-3	1	0	0	0
T	1	-1	0	0	-2	-1	0	0	-1	0	-2	0	-1	-3	0	1	3	-5	-3	0	0	-1	0
W	-6	2	-4	-7	-8	-5	-7	-7	-3	-5	-2	-3	-4	0	-6	-2	-5	17	0	-6	-5	-6	0
Y	-3	-4	-2	-4	0	-4	-4	-5	0	-1	-1	-4	-2	7	-5	-3	-3	0	10	-2	-3	-4	0
V	0	-2	-2	-2	-2	-2	-2	-1	-2	4	2	-2	2	-1	-1	-1	0	-6	-2	4	-2	-2	0
B	0	-1	2	3	-4	1	2	0	1	-2	3	1	-2	-5	-1	0	0	-5	-3	-2	2	2	0
Z	0	0	1	3	-5	3	3	-1	2	-2	-3	0	-2	-5	0	0	-1	-6	-4	-2	2	3	0
X	0	0	0	0	0	0	0	0	0	0	0	0	0	0	0	0	0	0	0	0	0	0	0

Figure 2.3. Mutation substitution matrix. The values represent the likelihood of each residue type being replaced by each other residue type in the process of evolution. (Source: Gray, 1990, p.12).

2.2. THE IMPORTANCE OF SEQUENCE DATABANKS IN SEQUENCE ANALYSIS

As mentioned above, one example of sequence database use is provided by the problem of finding coding regions in nucleic acid sequences. Indeed the value of searching a large sequence database becomes apparent when more and more homologues are found, revealing previously unsuspected relationships. For example Doolittle writes :

> "the real surprise came when a routine computer search revealed that the amino acid sequence of δ -crystallin from bird lenses and the enzyme argininosuccinate lyase are more than 55 percent identical" (Doolittle, 1988).

After listing some similarities (Doolittle, 1988, p.18) he goes on

> "even more surprising, a computer search of published sequence fragment from a squid lens crystallin suggested that this protein might be the enzyme glutathione S-transferase. It is this suggestion that Tomarev and Zinovieva (1988) confirm in their paper."

Similarly, research on heat shock proteins and proteins of the rough endoplasmic reticulum - which is part of a cell - led to the identification of one of the first DNA sequence elements shown to be responsible for gene regulation in eukaryotes. The sequence was proved to be the binding site for a specific heat shock transcription factor and had a common sequence with other proteins such as hsp 70 and grp 94 (Anon, 1989).

These examples demonstrate clearly the role and importance of sequence databases to molecular scientists. They have become research tools in the same way that computers have. They are crucially important in many areas of biochemistry, molecular biology and genetics.

2.2.1. Proposed second-generation databanks

The interpretation of data obtained from databanks (just like from other sources) is an important consideration. The knowledge of what is available and how it is accessed (which will be dealt with in later chapters) is only part of the manifold aspects of the uses of databanks. A suggestion of a more intensive potential future use will be mentioned where a need is addressed for "databases which would emphasise concepts and relationships" (Pabo, 1987).

One example of a higher level of potential future use is the possibility of gaining information from different sources on different levels. A future database of protein-DNA interactions for example would contain the amino acid sequences of DNA-binding proteins and the DNA sequences of their binding sites. The different levels of the mutations (changes of bases) that affect the protein and even the atomic details of the interactions, would all be relevant and useful.

Of even more radical a nature is the suggestion that the successful databases of the future will be fundamentally different from existing ones. It has been suggested that they will distinguish between scientific findings and the genetic information directly stored within cells by the organisms themselves. These two types of information are often confused and overlap in most existing databases.

2.3. INTEGRATION OF DATABANK SEARCHING WITH SEQUENCE DETERMINATION

With the increase in computing power of personal computers and with the introduction of sequence analysis software for IBM-compatible PCs, sequence determination and database search have been brought closer together. Software is now available on floppy disks or CD-ROM for systems such as DNASIS, DNASTAR, IBI/PUSTELL, PC/GENE, and MICROGENIE, which use sequences from GenBank, PIR, SWISS-PROT,and EMBL databanks (see chapter 7) and which have comprehensive data management systems with features such as

(1) data entry and editing,

(2) DNA and protein analysis,

(3) restriction enzyme site analysis and mapping, and

(4) sequence comparison and search for homologues.

Such software is convenient to use because it is in the laboratory. However, there are disadvantages. For example the length of sequence held on floppy disks is usually limited to 30-60,000 nucleic acid base pairs (equivalent to 30-60 kilobytes of computer memory). However recently floppy disks of much larger capacity (1.4 megabytes as at 1990) have become available. Another recent development is that databanks have become available on CD-ROM (Claverie, 1988). Also the data is updated (or the CD-ROM disks are replaced) less often than the frequent updates of online databanks. A detailed comparison of four such integrated laboratory software systems can be found in Hoyle (1987).

2.4. REFERENCES

Anon, 1989, "Grapevine: The EMBO medal", Trends Biochem. Sci., 14, 468-469.

Ausubel F.M., 1989, Current protocols in molecular biology, 1, 7.1.2.

Bach R., Friedland P., Brutlag D.L., and Kedes L., 1982, "MAXIMIZE : A DNA sequencing strategy advisor", Nucleic Acids Res., 19, 295-304.

Claverie J-M., 1988, Computer access to sequence databanks, 85-99 in Lesk A.M., (Ed.), 1988, Computational molecular biology: sources and methods for sequence analysis, Oxford University Press, Oxford.

Doolittle R.F.,1988, "Lens proteins, more molecular opportunism", Nature, 336, 18-19.

Friedland P., Kedes L., Brutlag D., Iwasaki Y., and Bach R., 1982, "GENESIS, a knowledge-based genetic engineering simulation system for representation of genetic data and experiment planning", Nucleic Acids Res., 10, 323-340.

Gingeras T.R.,Milazzo J.P., and Roberts R.J., 1978, "A computer assisted method for the determination of restriction enzyme recognition sites", Nucleic Acids Res., 5, 4105-4127.

Gingeras T.R., 1983, Computers and DNA sequences : a natural combination, 15-43 in Weir B.S., (Ed) Statistical analysis of DNA sequence data, Marcel Dekker, New York.

Gray N., 1990, "A program to find regions of similarity between homologous protein sequences using dot-matrix analysis", Journal of Molecular Graphics, 8, March ,11-15.

Hawley D.K., and McClure W.R., 1983, Nucleic Acids Research, 11, 2237.

Hoyle P., 1987, Use of commercial software on IBM personal computers, 47-83 in Bishop M.J., and Rawlings C.J., (Eds) Nucleic acid and protein sequence analysis : a practical approach, IRL Press, Oxford.

Needleman S.B., and Wunch C.D., 1970,"A general method applicable to the search for similarities in the amino acid sequence of two proteins", J.Mol.Biol., 48, 443-453.

Pabo C.O., 1987, Nature, 327, 467.

Pearson W.R.,1982, "Automatic construction of restriction site maps", Nucleic Acids Res., 10, 217-227.

Sellers P.H., 1974, "On the theory and computation of evolutionary distances", S.I.A.M. J.Appl.Math., 26, 787-793.

Sellers P.H., 1979, "Pattern recognition in genetic sequences", Proc.Natl.Acad.Sci.USA., 76, 3041.

Staden R.,1977, "Sequence data-handling by computer", Nucleic Acids Res., 4, 4037-4051.

Stormo G.D., 1987, Identifying coding sequences, 231-258 in Bishop M.J. and Rawlings C.J., (Eds.), Nucleic acid and protein sequence analysis: a practical approach, I.R.L. Press, Oxford.

Tomarev S.I.,and Zinovieva R.D., 1988, "Squid major lens polypeptides are homologous to glutathione S-transferases subunits", Nature, 336, 86-88.

von Heijne G., 1987, Sequence analysis in molecular biology, Academic Press, New York.

Waterman M.C., Smith T.F., and Beyer W.A., 1976,"Some biological sequence metrics", Adv.Math., 20, 367-387.

Chapter 3:

Online Databases in Biochemistry and Molecular Science

In the previous chapter computer applications in the laboratory were discussed giving sequence determination and analysis as examples. Here the importance of online facilities as information retrieval tools will be examined in some detail. The focus will be on the nature of information and on the need for up-to-date information which is easily accessible. Although the aim of this discussion is to draw attention to databases devoted to protein research a short review of online bibliographic databanks and databank products relevant to biochemistry and molecular science is provided and gives references for more information elsewhere.

3.1. THE IMPORTANCE OF ONLINE DATABASES

Maizell (1987) states that one of the most important developments in chemical information in the last ten years has been the growth and improvement of online databases. The size and intricacy of the subject, together with its rapid growth, mean that biochemists within even small specialisms cannot keep track of information using hard copy and manual search. For example there are over nine million distinct chemical compounds listed in Chemical Abstracts Service CAS Online. Comparison and retrieval of such information can only be done by computer search programs, which help the molecular scientist to comb online databases. Among the developments of the last decade, Maizell lists five as particularly important:

> "(1) The newer databases permit structure and substructure searching by both graphic and nongraphic methods. Structure searching was possible before, but not with nearly as much power as is available now.
> (2) Much more data about patents, especially United States patents, are available online now.
> (3) Full text online databases are more widely available.
> (4) Special attention is being given to facilitate access to databases through personal computers.
> (5) Numerical values (hard data) are increasingly available online "(Maizell 1987 p.152).

Further developments such as gateways, front-ends and networking could also be added. The arrival of easy-to-use gateways has enabled users to search the databases of several different hosts. They simplify the process of signing on and giving codewords, so that once one has entered the gateway one is effectively able to access the databases of several different hosts. In the molecular science area, Scimate and ProSearch (Kasselman and Watstein, 1988, Ch.10) are important examples of gateways. Several programs exist which make it easier for the user to develop search strategies based on different criteria and rules. STN Express is an example of such a front-end program which can be used on PCs for constructing a search strategy off-line (i.e. before the connection is made to the online host's database). Networks are telecommunication or cable links which allow direct communication between software systems (allowing sharing of databases among other things) and hardware systems (allowing sharing of computing power). An important example of a network for molecular biologists is BIONET, which is provided by Intelligenetics Inc. It provides scientists with access to many databases (including GenBank, EMBL, PIR, RED, the Brookhaven Protein Structure Database, and VectorBank) and also to search and manipulation software (Brutlag and Kristofferson, 1989).

3.2. WHY USE ONLINE SERVICES?

Why use online services? Online databases have entry points which allow much faster and more accurate search than is possible in printed materials. These entry points comprise authors, keywords, formulas and patent numbers as well as some that are unavailable in printed materials, such as chemical substance names, atomic weights, CAS registry numbers, chemical notations, graphic structures, periodic groups, substance classes and generic features.

Online services allow users to search according to Boolean operators for example

A or B or C or (D and E) or F or G

They also allow searchers to specify the type of publication wanted, language, or the country of publication. However, for many online databases coverage does not go back beyond the 1960s. For example while the printed Chemical Abstracts started in 1907, the online version of CA's bibliographic files does not start until 1967. So while online services enable the searcher to search with certainty, this depends upon the quality and completeness of the original sources. Another problem is that molecular information, both bibliographic and numeric, is often registered and indexed in an inexact way. Studies described later found that substances were wrongly classified on many occasions. So a thorough knowledge of the database being searched is as essential for online searching as it is for manual searching (and the use of more than one online database is nearly always to be recommended).

Molecular scientists express their ideas in graphical terms, on pieces of paper usually, when they wish to investigate an idea or communicate with other molecular scientists. It is therefore of the first importance that information can be input, output, manipulated, searched for, and saved, in graphical terms. This has been driven by the online industry's desire since 1980 to compete with the graphics capabilities of modern and powerful PCs. But for molecular scientists the implications are vast. Graphic searching provides freedom from nomenclature substitutes, such as classification and coding systems. All of these (nomenclature, classification, and coding systems) have a certain amount of artificiality and are difficult to learn. In addition, graphic searching provides the potential for more effective searching for less well defined structures, including true generic searching. For example, parts of a structure can be marked as fixed and substructures incrementally altered as in Markush structures in patent searching : this is easier to do as a graphics representation than as a linear formular or name. Also many aspects of molecular modelling become easier with graphics, since shape and position within a structure are so important. Many organic substances generate problems about how to alter shape and structure to enable new structures to fit around and inside them, thus generating new substances with new properties. Thus graphic searching is consistent with the work habits of most molecular scientists. The graphics systems at present available allow structures to be drawn wery much as molecular scientists have traditionally drawn them or alternatively in one of the linear notations or by means of coordinate pairs. This last method is most unusual now that quick and easy direct drawing input is available.

3.3. WHAT PROBLEMS MOTIVATE USING ONLINE DATABASES?

What are the principal types of problem which molecular scientists wish to solve that motivate them to use online databases?

(1) Firstly molecular scientists working on proteins wish to know when they find a new protein. This involved comparing with a database of proteins.

(2) Secondly they wish to discover the protein's three-dimensional structure. This is a very difficult procedure which will be described in later chapters. It involved consulting databases at a number of points in the process.

(3) Also they wish to model molecules, by using similar molecules whose 3D characteristics are better known than the molecule to be modelled. Again this will involve a search of a database.

(4) Also it is desirable that they should avoid "discovering" a compound or process which someone has already found or patented. The problem here is to search for a patent which answers the scientist's problem (perhaps a compound or process which has a set of desired attributes). This may be difficult since although scientists know what the problem looks like, they will not yet know what the solution looks like. The second kind of problem related to patents is when a genuine discovery has been made. The scientist then needs to describe his discovery in as general a way as possible, so that the licence fee can be maximised, but not so general that the patent strays onto someone else's already patented territory. The search problem here is one of successively widening or narrowing (in particular ways) the scope of definition of the subject of search.

(5) Another way in which scientists use databases is for "browsing". This can be defined as the extending of specialised knowledge in the absence of a complete specification of what is interesting. This browsing problem is currently receiving much attention.

Several approaches, all of them involving approximation methods, are being pursued. The results are promising : "browsing" seems to be one of the things that computers can do well. The recent advent of graphics has led to many algorithms which find approximations to the sought structure or fragment, or inversely which have specified differences essentially by means of pattern recognition.

(6) Another type of problem is concerned with finding how to synthesise a (known) product compound from a (known) educt compound via an (unknown) pathway. Here the intermediate steps have hints, such as bibliographic references to intermediate compounds and their properties, but usually a great deal of biochemical knowledge is required in guessing where the pathways lie. Although much research is being devoted to this area of reaction pathway retrieval, no completely successful algorithm has yet been devised.

Besides the large number of chemical compounds that have so far been discovered, there has also been a fast growth of publications in all fields of science and technology especially since the war. The Science Citation Index (SCI) contains about 15 million items published between 1945 and 1989 with over 175 million cited references (Garfield, 1990). As a proportion of all documents in Chemical Abstracts, the biochemistry section, for example, represents 26-34 percent of the papers published between 1968 and 1977 (Rowland, 1984). There are other calculations which are slightly different (Turner, 1987). It seems that between 1975 and 1984 the number of documents was nearly the same : between 150,000 and 162,000, which is 34-35 percent of the total abstracts in all sections of the Chemical Abstracts. The latest surveys of trends on the biochemistry literature (Garfield, 1979, Bottle and Gong, 1987) show that the biochemistry literature is still growing in absolute and relative terms although slower than before.

If we have a look at the list of the 100 most cited papers in the SCI between 1945 and 1988 (see table in Garfield, 1990), it turns out that most of them describe methods for protein or amino acid analysis. The top two are Lowry's (1951) paper with over 180,000 citations and Laemmli's (1970) publication with over 60,000 citations, while the two most often cited recent ones were Sanger and Coulson's paper (1977) with over 10,000 citations and Maxam and Gilbert's publication (1980) with nearly 9,000 citations. It can be seen from these studies as well that the number of publications in this field is rather high and that it would be very time consuming to do retrospective searches without using online facilities.

3.4. TYPES OF ONLINE DATABASES AND CD-ROMS

There are three types of database :

(1) Bibliographic (references, abstracts). Example of bibliographic databases are MEDLINE, BIOSISP (BIOSIS Previews) and TERRE-TOX. Both BIOSIS and MEDLINE have over 5 million entries, including citations and abstracts. BIOSIS Previews reports on the whole of the life sciences, while MEDLINE (and its retrieval system MEDLARS) deals with biomedicine and molecular biology. The much smaller TERRE-TOX database (Meyers and Schiller, 1986) of 15,000 studies dealing with acute toxicity, behaviour, reproduction, physiological, and biochemical processes. The TERRE-TOX studies have been abstracted into 6 predetermined "fields". For example,

the second field contains information on the role of the chemical investigated: whether it is an active ingredient, a test substance, a synonym of a tested chemical,a molecular formula, and so on. These fields predetermine how and with what result a search can take place.

(2) Numeric- , formulae-, or graphics-based databanks, which include some text. Examples of this type of databank are the sequence databanks or the genetic map databanks. The Canadian Scientific Numeric Data Service CAN/SND databank service, for example, provides online access to databanks in crystallography, molecular biology, spectroscopy, and chemical thermodynamics (Wood et al, 1989).

(3) Full text databases. About 30 journals (some of them biomolecular journals) are available online in full text. Any word, string of words, or alphanumeric term can be searched for.

Most of these databases are available on a number of hosts such as Data-Star, Dialog, BRS, Pergamon Orbit Infoline, STN, CIS, BLAISE-LINK, just to mention the main ones in molecular and biochemical science. The lack of standardisation and the variety of access protocols and command languages used are some of the problems which may limit their use. For more information on online searching bibliographic guides can be consulted such as those by Byerly (1983) and Batt (1988). These and the subject guides (e.g. Morton and Godbolt 1984, Parker and Turley, 1986; Wyatt, 1987) list and describe them by discipline rather than dividing them into bibliographic and non-bibliographic ones. There are online directories (e.g. Lawton et al, 1989, and Cuadra/Elsevier, 1990), and printed ones which list them either alphabetically (Stephens, 1986) or by subject. Erdmann et al., 1989, p.320-323) provide a comprehensive list of the main online databases in the life sciences.

The number of numeric-,formulae-, or graphics-based databases (sometimes called databanks - see chapter 4 for a definition) is growing in science and technology. These will be described in some detail in the following chapters for molecular bilogy, genetics, and biochemistry and will be referred to as sequence databanks and structure databanks. These are mainly those which are used by scientists who are involved in genetics or protein sequence and structure analysis. Review articles on other numeric databanks should be consulted such as the one by Allan and Ferrell (1989) or some others in journals such as Database, Online Review, Journal of Information Science, Aslib Information, Information Processing Management, Journal of Documentation, etc. The online version of well-known handbooks, (e.g. Beilstein, Merck Index, Fine Chemicals Handbook) are now available on different hosts (Merck Index on BRS and DIALOG; Beilstein on DIALOG and STN).These handbooks will probably soon become available on CD-ROM as have other databases in the past including Chem-Bank, Medline, Excerpta Medica, Agricola, Biological and Agricultural Index, Compact Aquatic Sciences and Fisheries, among others of relevance to molecular and biochemical science. They are all listed in CD-ROMs in Print compiled by Emard (1989). Some guides (Armstrong and Large, 1990) and recent articles (Miller, 1987; Large, 1989; Robertson, 1989) can be recommended where the authors evaluate and compare CD-ROM products. There is a growing number of user studies and most of them indicate that CD-ROM is popular, mainly in academic libraries (Fox, 1988; Poisson, 1989). Fox reported that

> "the heaviest use was of end-user searching, and 83 percent of the libraries with CD-ROM reported having at least one product for this purpose" (Fox, 1988).

Some bioscience databanks are not found on any of the major online hosts, but are available through networks or can be purchased on magnetic tape or floppy disk. Networks such as the Joint Academic Network (JANET) which links academic users in the UK with many foriegn hosts (for example, GOS or GenBank Online Service; the National Network of the GenBank System; the BITNET network system in the U.S.A., or networks in Europe and Japan). Joint projects, oordination of data acquisition and management, and standards in format, are headed by CODATA, which has a publication on sources and trends (CODATA, 1982).

The increase in power and variety of the technology involved has led to a gradual merging of the various components of the database environment. There are powerful forces motivating large science-based companies to instal their own databanks and databases. These forces include confidentiality and security, the requirement to keep methodical housekeeping results (both positive and negative results), and methodical records of transportation and the dating of patents. Yet at the same time such organisations are becoming more and more dependent upon online services. These two information sources are, however, essentially complementary. The biochemist may wish to input a structure from an inhouse database and search online for whether it has been patented.

Such a merging is taking place in many ways. There has been a recent development towards dual-mode text-or-graphics capabilities with the advent of good graphics facilities on PCs. This capability has permeated all molecular-related software. Also the large expansion of memory of PCs has led to software for uploading previously-prepared search queries on to online services and for the inverse operation thus making it possible for the molecular scientist and biochemist at his PC to have at his fingertips software and information prepared for PC, for CD-ROM, for internal databases, or for online access. Indeed, software designed initially for online use has reached the PC and CD-ROM market in slightly altered but not less powerful form. Software, capable of dual mode (graphics or text) and increasingly intelligent, is being created for molecular scientists and biochemists in the following areas : data handling, molecular information systems, tools for theoretical molecular science and biochemistry, molecular modelling, statistical analysis packages, and report generators (with graphics and laser-printer capability). Also housekeeping packages exist for transportation records, patent dating, and experimental result logging. So integration is becoming a currently important issue, allowing the laboratory scientist to switch from retrieval to modelling to experiment and back to retrieval again.

3.5. THE FINANCING OF DATABASES

Some databases are purely financed by government, while others are totally funded by private enerprise, and some are a mixture of the two. This mix has arisen without too much care or worry, since it is easy to argue both that some databases can only be viable if supported by public money and also that private companies are best at finding customers for new products and services.

This situation is likely to be disturbed somewhat by a legal action taking place in the US. The action is between the American Chemical Society (ACS) whose Chemical Abstracts Service (CAS) provides the largest single chemical research data source in the world on a not-for-profit basis and Dialog, one of the biggest sellers of information, including CAS data, worldwide. Of the 100 million dollars of business that CAS does each year, the largest individual share involves customers buying CAS data from Dialog. However, the importance of the current lawsuit is shown by the fact that Dialog could get 150 million dollars from CAS if it can prove that CAS is a commercial, monopolistic enterprise which abuses its monopoly by favouring some of its customers, usually the larger ones, over others.

The conflict has been sparked off by the fact that the ACS has decided to include molecular sequence and structure searching as well as the standard abstract-searching features, but that such additional facilities will not be available for users of CAS via Dialog. Dialog is arguing that this puts the ACS in direct commercial competition with Dialog, and this is unfair because the ACS enjoys tax exemptions reserved for benevolent not-for-profit organisations.

Other users of CAS data, such as those using the Royal Society of Chemistry service in Britain, are able to enjoy the full range of ACS facilities.

So far database development has been relatively free of litigation. And studies by the European Commission show that the smaller EC countries are more likely to have a higher proportion of not-for-profit databases. For example, Spain has 77 not-for-profit compared with only 10 profit-making databases. The UK on the other hand, has 92 not-for-profit and 194 profit-making databases, reflecting the lack of government concern about support and direction. In the US, there are 314 not-for-profit and 1532 profit-making databases, although the higher research funding from government in the US has meant that many commercially-oriented database companies have grown up on the back of government contracts. For example, Dialog in the US grew as a result of contracts from the National Aeronautics and Space Administration. In Europe, there are some similarities with the European Space Agency, which has a division called ESPRIN which supports ESA databases. Presumably the European Commission's patronage for databases will expand into other areas in the future, including molecular databases.

3.6. TRAINING END-USERS

The price of these new developments has been that scientists have had to become more expert end-users. It used to be true that molecular scientists and biochemists were reluctant online users, relying on expert intermediaries. This is less true today, when only the more difficult searches, including patent searches, are entrusted to intermediaries rather than to end-users, ie the scientists who actually require the information. There has also been an increased interest in evaluation of information systems in terms of their usefulness and accessibility to different types of users (Bawden, 1990). Major companies have sought to train their scientific personnel, and scientists are becoming used to computers in the laboratory anyway. Any reluctance on the part of librarians and information scientists to allow molecular scientists and biochemists to do their online searches directly is probably misplaced. Buntrock and Valicenti (1985, p.203) comment

> "End user searching is inevitable. Participate in the training of end-users or, by inaction or, worse yet, by active resistance, participate in your own demise".

Studies of end-user training by ICI (Warr and Haygarth Jackson, 1988), by DIALOG (Palma and Sullivan, 1988), and by Amoco (Buntrock and Valicenti, 1985), generally reported favourable results of training sessions. Warr and Haygarth Jackson (1988) suggest that

> "Information scientists are invaluable for end-user support for supplying information on system updates, and for supplying user-oriented documentation . The chemists realised that the information scientists had considerable expertise, and as a result so far as chemists were concerned the status of the information scientists increased. Furthermore,the improved knowledge of online searching acquired by chemists helped in query definition for complex online searches, and an appreciation of the capabilities of online searching produced

> more interesting and challenging questions for information scientists. The number of information scientists has increased due to the increased visibility of the importance of online searching", (Warr and Haygarth Jackson, 1988, p.70).

It is difficult to tell how widespread end-user searching is in science and industry. This is partly because scientists and technologists do not often write for an information science audience. Also the case studies tend to be of the industry leaders who tend to be more advanced than other companies. Also, the evidence from the U.S.A. has a similar bias; probably end-user searching is more advanced there (by about 3 years according to Nicholas *et al.*, 1988). Also studies using information supplied by hosts (for example Data-Star by Nicholas *et al.*, 1988) tend to paint a rosy picture of end-user searching since only the enthusiasts are studied, rather than those who are reluctant to become end-users.

3.7. CURRENT AWARENESS SERVICES AND IN-HOUSE SYSTEMS

Current awareness is the process of keeping up-to-date with information on recent developments in the subject field. There are at least three ways of keeping up-to-date :

(1) scanning primary or abstracting journals,

(2) scanning current awareness publications (e.g. the weekly issues of Current Contents (Life Sciences), CA Selects, BIOSIS CONNECTION, CABS and so on,

(3) using the automatic SDI facility of online databases.

The third method will now be explained in some detail together with a survey of some of the latest developments. There is currently a growing need to download and reformat information from online databases and to build a comprehensive database to fulfil particular user needs. Warr and Haygarth Jackson (1988) state :

> "To the ICI end-user who can access ICI's SAPPHIRE database in England and CAS databases through Karlsruhe using the same IBM-PC, the distinction between internal and external databases is somewhat artificial".

The same can be said about a molecular biologist who uses GenBank on floppy disk or CD-ROM using the same program as the main online systems for sequence homology searches. Indeed, the trends are towards integrated systems. Workstations are suitable for carrying out different yet integrated tasks which are carried out in research laboratories.

Similarly, there is a need for a subject-oriented (or rather, need-oriented) current awareness bulletin as was recognised by a few companies some years ago (Cosgrave and Wade 1980). This can be done using the automatic SDI facility of online databases. The method is based on a few search profiles with broader search strategies and is flexible enough to respond to changing needs and make sure that no relevant records are missed. Because the number of references and records between two updates is quite small, compared with the standard retrospective search, these searches are fast and cheap. Commands exist to save the strategy (Parker and Turley 1986) and the host computer will execute it each time the database is updated. The results can be printed out and mailed or downloaded to the scientist's PC. A new service marketed directly to end-users is BIOSIS Connection (BC). This is targetted to serve the current awareness needs of life scientists. In May 1988 it included six databases BIOEXPRESS, BIOMEETINGS, BIOBOOKS, BIOPATENTS, FORTHCOMING EVENTS, SERIAL SOURCES FOR BIOSIS DATABASE, but more were added later: BIOTHESES, JOBLINE, AIDS IN FOCUS.

BC does not index or abstract citations while BIOSIS Previews (BP) does. BIOSIS together with B-I-T-S (BIOSIS Information Transfer System, which is a prearranged SDI profile executed monthly on BP BIOSUPERFILE II), have been compared and the possibilities of building a searchable local retrospective database by scientists have been examined by Brueggeman (1988a). Recently another current awareness service on floppy disk was announced: the Medical Science Weekly. It is searchable on menu or by command strings with Boolean operators and is suitable for IBM-PC and compatibles. It is aimed at end-users in the academic biosciences and is based on the contents pages of medical and biomedical journals (see Aslib Information, 1990).

3.8. REFERENCES

Allen F.C., and Ferrell W.R., 1989, "Numeric databases in science and technology: an overview", Database, 121, 50-58.

Armstrong C.J., and Large J.A., (Eds) , 1990, CD-ROM information products : an evaluative guide and directory, Gower, Aldershot.

Aslib Information, 1990, January.

Batt F., 1988, Online searching for end-users : an information book, Oryx.

Bawden D., 1990, User-oriented evaluation of information systems and services, Gower, Brookfield, Vermont, USA.

Bottle R.T., and Gong Y.T., 1987,"A bibliometric study on the ageing and content typology relationship of the biochemical literature", Journal of Information Science, 13, 59-63.

Brueggeman P.L., 1988a,"BIOSIS Connection : linking end users with current awareness information", Database Searcher, 4, 24-30.

Brutlag D.L., and Kristofferson D., BIONET: an NIH computer resource for molecular biology, 287-293 in Colwell R.R., Biomolecular data: a resource in transition, Oxford University Press, Oxford. sp Buntrock R.E., and Valicenti A.K., 1985,"End users and chemical information", J.Chem.Inf.Comput.Sci., 25, 203-207.

Byerly G., 1983, Online searching : a dictionary and bibliographic guide, Libraries Unlimited.

CODATA, 1982, Inventory of data sources in science and technology : a preliminary survey, UNESCO, Paris.

Cosgrave R.G.M., and Wade L.G., 1980, "users' reactions to a corporate-designed current awareness bulletin", J.Chem.Inf.Comput.Sci., 20, 179-181.

Cuadra/Elsevier, 1990, Directory of online databases, Elsevier.

Emard J.P., 1989, CD-ROMs in print 1988-89, Meckler.

Erdmann V.A., Klussmann U., Wolters J., and Beckmann H.-O., 1989, Computer education in biochemistry, chemistry, and molecular biology (II), 317-325 in Colwell R.R., Biomolecular data: aresource in transition, Oxford University Press, Oxford.

Fox D., 1988, "CD-ROM use in Canadian libraries : a survey", CD-ROM Librarian, 3, 23-30.

Garfield E., 1979, Trends in Biochemical Sciences, 4, 290-295.

Garfield E., 1990, "The most-cited papers of all time SCI 1945-1988. Part 1A. The SCI top 100 - will the Lowry method ever be obliterated?", Current Contents No.7, 3-14.

Kasselman M., and Watstein S.B., 1988, End user searching : Services and providers, A.L.A.

Laemmli U.K., 1970, "Cleavage of structural proteins during the assembly of the head of bacteriophage T4", Nature 227, 680-685.

Large J.A., 1989,"Evaluating online and CD-ROM reference sources", J.Librarianship 21, 87-108.

Lawton J.R., Martinez F.A., and Burks C., 1989, "Overview of the LIMB database", Nucleic Acids Research, 17, 5885-5899.

Lowry O.H., Raseborough N.J., Farr A.L., and Randall R.J., 1951,"Protein measurement with the Folin phenol reagent", J.Biol.Chem. 23, 265-275.

Maizell R.E., 1987, How to find chemical information : a guide for practicing chemists, educators and students, 2nd Ed., Wiley Interscience, (10), 152-200.

Maxam A.M., and Gilbert W., 1980,"Sequencing end-labelled DNA with base-specific chemical cleavages", Meth. Enzymology, 65, 499-560.

Meyers S.M. and Schiller S.R., 1986, "TERRE-TOX : a database for effects of anthropogenic substances on terrestrial animals", J.Chem.Inf. Comput.Sci., 26, 33-36.

Miller D.C., 1987, "Evaluating CD-ROMs : to buy or what to buy?", Database 10, 36-42.

Morton L.T., and Godbolt S., (Eds), 1984, Information sources in the medical sciences, 2nd Edition, Butterworth.

Nicholas D., Erbach G., Pang Y.W., and Paalman K., 1988, End Users of Online Information Systems : an Analysis, Cambridge University Press, Cambridge.

Palma M.A., and Sullivan C., 1988, "Meeting the needs of the end user", J.Chem.Inf.Comput.Sci., 25, 422-425.

Parker C.C.,and Turley R.V., 1986, Information sources in science and technology, Butterworth.

Poisson E.H., 1989, "CD-ROM in biomedical libraries : a survey of implementation", CD-ROM Librarian, 4, 13-16.

Robertson C., 1989, "Silver Platter's CHEM-BANK", CD-ROM Librarian, 3, 30-34.

Rowland J.F.B., 1984, Biochemistry, biophysics, and molecular biology, 142-173 in Morton L.T., and Godbolt S., Information sources in the medical sciences, Butterworth.

Sanger F., Nicklen S., and Coulson W., 1977, "DNA sequencing with chain-terminating inhibitors", Proc. Nat. Acad. Sci. U.S.A., 74, 5463-5467.

Stephens J., 1986, Inventory of abstracting and indexing services produced in the U.K., B.L.

Turner J.M., 1987, Biomedical sciences, 57-80 in Wyatt (Ed.) , Information Sources in the life sciences, Butterworth, 1987.

Warr W.A., and Haygarth Jackson A.R., 1988, "End user searching of CAS Online : results of a cooperative experiment between Imperial Chemical Industries and Chemical Abstracts Service ", J.Chem.Inf. Comput.Sci., 28, 68-72.

Wood G., Rogers J.R., and Gough S.R., 1986, "Canadian Scientific Numeric Database Service", J.Chem.Inf.Comput.Sci., 118-123.

Wyatt R.V., (Ed.), 1987, Information sources in the life sciences, Butterworths.

Chapter 4:

Methods for Computer Representation and Registration

In order to express information about chemical compounds the two dimensional structure diagram has proved to be sufficient and can be used for the purpose of identification and documentation in many cases. In fact chemistry is unique among the sciences in having such an informative and succinct formal language. In the case of modern molecular biology, such information is usually sufficient but in some cases must be supplemented by three-dimensional information before many of the biologically significant data can be extracted.

This chapter describes some of the ways in which molecular structures are represented in a computer. It also describes the problem of uniquely identifying each item (the problem of registration), and discusses the role of the registry file within the molecular information system.

4.1. AMBIGUOUS VERSUS UNAMBIGUOUS REPRESENTATION

A characterisation of a molecular structure is ambiguous if it represents more than a single structure, or unambiguous if it represents a single structure only.

4.1.1. Ambiguous representation

Usually, ambiguous representations provide only partial information about a molecule. The molecule is described in such representations by a collection of its characteristic fragments. Reconstruction of the complete molecule from such fragments is not usually possible. However, ambiguous representations have an important classificatory function and are much used in indexing and organising collections of chemical compounds. So a fragmentation code is designed with a particular collection in mind. If the compound collection is poorly categorized by the fragmentation code, then the effectiveness of any derived indexes is also poor. For this reason fragmentation codes are collection-specific and are not usually portable to other collections.

For example, the GREMAS code (Rossler and Kolb, 1970) constructs fragment terms according to broad structural classifications, each of which is divided in a hierarchical manner. Each term is described by a three letter code. The first letter indicates the genus symbol, the second letter indicates the species (and shows whether the compound is a free acid or a specific derivative) and the third indicates the subspecies, i.e. the environment of that species (including designations for aliphatic, alicyclic, aromatic, and heterocyclic environments).

However, fragmentation codes are ambiguous. For example, the two structures in Figure 4.1 have the same fragment terms (NOA, EDA) yet comprise two distinct structures. The ambiguity must be removed by additional information, using a special grammar, to distinguish the two compounds before a search can be guaranteed to be successful (see Figure 4.1). Otherwise false drops will cloud the search, leading to much wasted time (but not to loss of relevance). This method has proved so successful that effective searching is possible using only the fragment description.

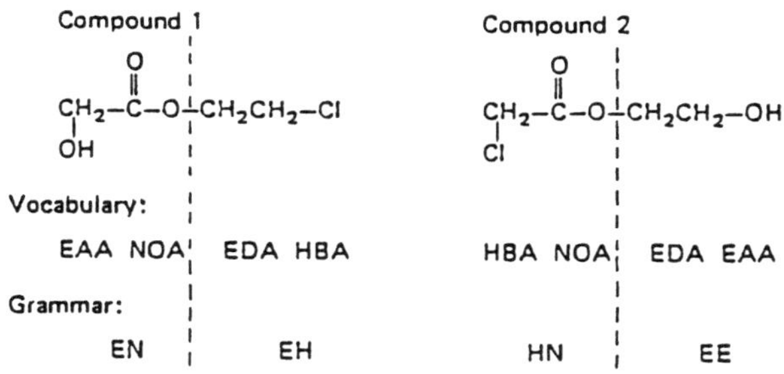

Figure 4.1. Two distinct compounds with identical GREMAS fragmentation code. (Source: Ash et al., 1985, p.134).

Although fragmentation codes are useful when incomplete information exists about a molecule, there are also situations which require that not all the information be included, or that some information is left purposely vague. Markush structures in patents are of this type, so that the ability to characterise a molecule at different levels of vagueness or concreteness (for an example see Figure 4.2) is a useful one to have.

4.1.2. Unambiguous representation

Systematic nomenclature, linear notations, and connection tables are all examples of unambiguous representations of molecular information. Each of these methods can unambiguously represent the connectivity of the complete molecule. The availability of automatic translators has rendered these representation forms equivalent in most cases, and has also led to the predominance of unambiguous over ambiguous representation methods.

Systematic nomenclatures are used for generating unique index names for each new substance name. Two are internationally recognised: the International Union of Pure and Applied Chemistry (IUPAC), and the Chemical Abstracts Service (CAS) nomenclature (Rowlett 1975). Automatic gen-

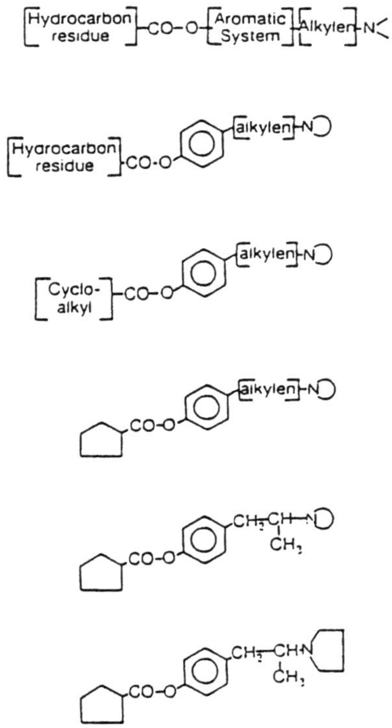

Figure 4.2. Generalisation of a specific molecule. (Source: Ash et al., 1985, p.136).

eration algorithms exist for systematic nomenclature (for example, Mockus et al., 1981).

Any representation can be either in terms of computer records of molecular information at each distinct co-ordinate pair, or in terms of the connectivity and structural relationships of the molecule. This latter approach, the topological one, is the one most widely used, and it is worthwhile spending some time outlining how it can be used to construct connection tables and linear notation schemes.

Figure 4.3 shows a structure diagram for a molecule together with its connection table. Each atom has a unique number. Connection columns indicate atom numbers to which connections exist. Bond orders (single,

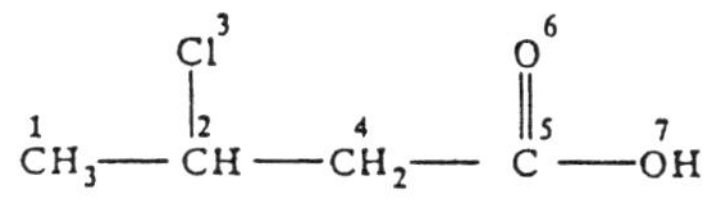

Atom No.	Atom Value	Con-nection	Bond	Con-nection	Bond	Con-nection	Bond
1	C	2	1	–	–	–	–
2	C	1	1	3	1	4	1
3	Cl	2	1	–	–	–	–
4	C	2	1	5	1	–	–
5	C	4	1	6	2	7	1
6	O	5	2	–	–	–	–
7	O	5	1	–	–	–	–

Figure 4.3. Connection Table for a simple acyclic structure. (Source: Lynch et al., 1971, p.14).

double, aromatic) between atoms are given by the bond columns. By convention hydrogen atoms are omitted, and are assumed to fill all unsatisfied valencies. Connection tables of the form of Figure 4.3 contain redundant information. A redundant connection table is most appropriate for the purpose of atom-by-atom searching, while a compacted form may be more desirable for storage of structure records in large files.

In the current CAS Registry (III) System the connection table is stored in non-redundant form. However, also all acyclic atoms are stored explicitly in a separate file and each ring is cited only as a ring identifier number (RIN) as in Figure 4.4. This reduces the storage required and speeds structure drawing and index name preparation (Dittmar et al., 1977, Vander Stouw et al., 1976). It is important to point out the necessity even in modern systems to save space. For example, the CAS Chemical Registry System stores chemical structure records whose size is restricted to 253 non-hydrogen atoms. Although nucleic acid and protein sequences appear in CA indexes with Registry Numbers

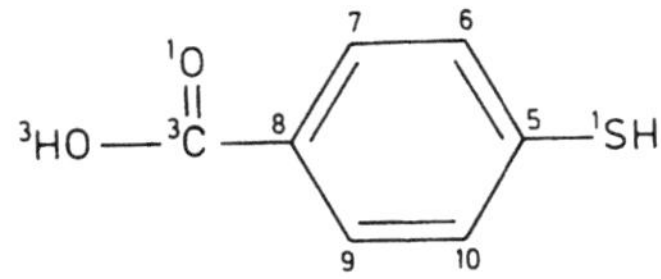

Registry III connection table consists of:

(a) Listing of ring identifiers

46T.150A.182 (RIN for benzene)

(b) Connection table of all acyclic atoms and bonds

	Connection	Bond
1 S	-	-
2 C	-	-
3 O	2	1
4 O	2	2

(c) Connections between ring(s) and acyclic components

1 - 5, 2 - 8

(d) Cross-reference for ring nodes between ring and substance numbering

1 2 3 4 5 6	Ring numbers
5 6 7 8 9 10	Substance numbers

Figure 4.4. CAS Registry (III) connection table. (Source: Ash et al., 1985, p.143). Reprinted with permission from Ash J.E., Chubb P.A., Ward S.E., Welford S.M., and Willett P., "Communications storage and retrieval of chemical information", Copyright (c) 1985, Ellis Horwood.

and names, they have structural records only if they have fewer than 13-15 nuceotides or 25-30 amino acids (Gleschen, 1989).

In the MACCS system (Molecular Access System) atom coordinates are stored in the connection table and enable the three-dimensional display of structures (Ash et al., 1985, p.182). The CROSSBOW system on the other hand, has connection tables which combine information about WLN (see below) and bond-explicit connectivity information (Ash and Hyde, 1975).

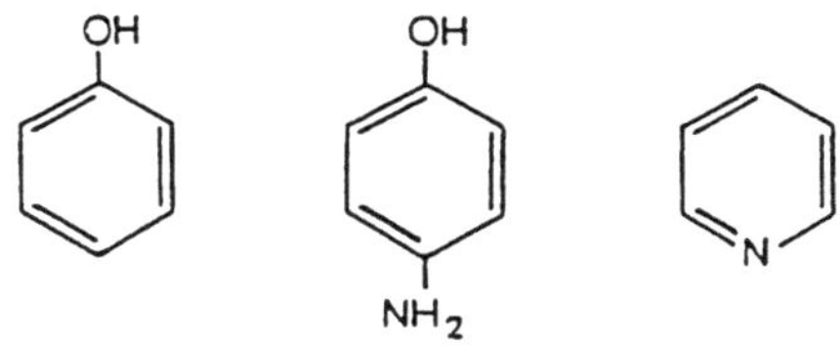

QR ZR DQ T6NJ

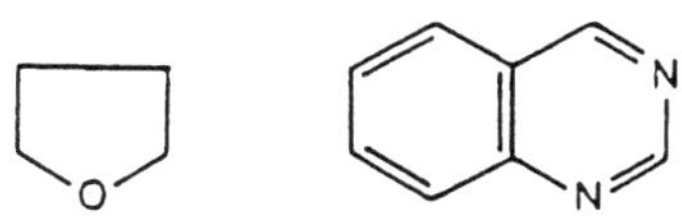

T5OTJ T66 BN DNJ

Figure 4.5. WLN ring notations.
(Source: Ash et al., 1985, p.140).
Reprinted with permission from Ash J.E., Chubb P.A., Ward S.E., Welford S.M., and Willett P., "Communications storage and retrieval of chemical information", Copyright (c) 1985, Ellis Horwood.

Besides connection tables, another unambiguous representation method is linear notation. Linear notations are relatively compact and economical. They usually describe a structure with fewer symbols than there are atoms. Also they contain information not contained in connection tables. One of the main linear notations is the Wiswesser Line Notation (WLN). The 40 WLN symbols represent particular atoms or functional groups. Thus the symbol Z represents a primary amino-group and the symbol 1 a methyl or methylene group. An alphabetical character preceeded by a space denotes the position of a substituent on a ring. The coding process proceeds by coding with a symbol each unit in the diagram, and writing each symbol in turn. So each atom is stated explicitly. The one exception to this is a ring structure (see Figure 4.5) where a generalised ring description is given, followed by an indication of ring size, the position of non-carbon atoms, the degree of unsaturation, and the means of fusion with other rings. Letter locants are used to indicate the positions of substituents on rings. Each locant is preceeded by a space in order to distinguish it as such, and is followed directly by the symbols for their respective substituent.

When a branched structure is described the coding process goes through three stages, from structure diagram to a diagram using WLN symbols, to a linear notation using WLN symbols. Double bonds are denoted by the symbol U, triple bonds by UU. Rings are described differently from chains, since the individual carbon atoms of rings are seldom explicitly cited. Benzene rings, because of their frequency, are given special treatment using the symbol R. Figure 4.6 shows three examples of linear notations and shows that WLN is simple and economical.

However none of these linear notations provides a unique, algorithmically-derived representation, i.e. one where the algorithm ensures one preferred coding for a given structure. This is not the case for connection tables and systematic nomenclatures, where algorithms exist for conversion between unique and non-unique representations. However,in 1989 an automated chemical structure descriptor called SMILES was developed (Weininger 1988, Weininger et al., 1989), which converts chemical structures into one, unique notation. The language uses the principles of molecular graph theory and allows rigorous specification using a very small and natural grammar. Although the line sequence which must be keyed in is longer than other linear notations, this is not important since once the string has been read into memory all further manipulation is automated. The notation is convertible into the other representation forms such as connection tables or other linear notations, so it can be used with any existing database. The notation uses the graph of a chemical structure, although it is not merely the

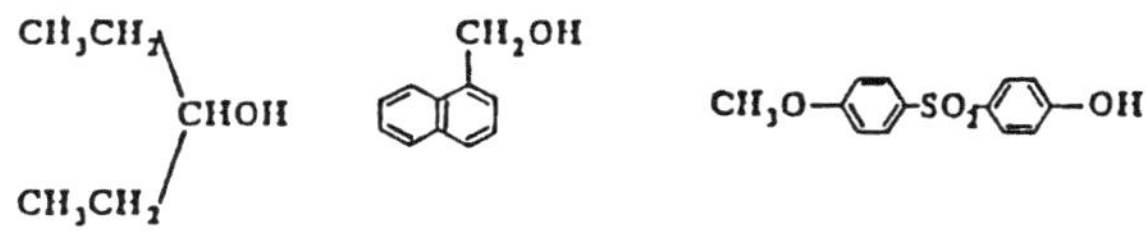

Hayward:	ECQE	6R(CQ)R3YR4Y	6R(SW@6RRRRQRR) RRRC(OM)RR
IUPAC:	C_5Q3	$B6_2CQ3$	$B6QC{:}4SO_2/4B6Q$
Wiswesser:	QY2&2	L66J B1Q	QR DSWR DO1

Figure 4.6. Three different linear notations.
(Source: Lynch et al., 1971, p.22).
Reprinted with permission from Lynch M.F., Harrison J.M., Town W.G., and Ash J.E., "Computer handling of chemical structure information", Copyright (c) 1971, Macdonald.

nodes (atoms) and edges (bonds) which are represented. In fact SMILES uses six different scale factors: atoms, bonds, branches, cyclic structures, disconnected structures and aromaticity.

4.2. GRAPHICAL DATA REPRESENTATION

4.2.1. Internal representation of graphical information

There are many different internal representation methods for graphics but just one will be discussed here as an example. It is Graphical Data Structures (GDS). It is used by Chemical Abstracts Service (CAS) for registration (i.e. unique naming to provide identifiers for unambiguous retrieval) of chemical substance information, together with connection tables. The GDS is used for structure input and display, and connection tables are used for structure searching. CAS has used GDS for its graphics- and text-database search programs Messenger and the PC-based front end to Messenger called STN Express. STN Express is at present the most sophisticated graphics package for searching large databases. It was released in Autumn 1988 and contains an offline structure input capability developed for structure query formulation.

GDS consists of 4 data blocks:

(1) Nodeblocks, which indicate hierarchy. In Figure 4.7, a node is shown by the symbol <node-type>.

(2) Branch blocks connect nodes to nodes or nodes to leaves. Each branch has a record containing the coordinates, intensity, scale and other parameters of the subtree below it. In Figure 4.7 it is represented by an arrow.

(3) Leaf blocks contain all the characters and lines and are shown in Figure 4.7 as <leaf-value>.

(4) Data blocks contain the application-dependent information such as atom and bond types and connectivity information. They are not shown graphically.

Figure 4.7 shows the GDS and connection table for caffeine (Haines, 1989, p.132). The order of attachment of a branch to a node is not significant. All the blocks are connected by circularly linked lists ("rings"). Each atom and bond leaf has a data block and these data blocks are all connected by rings. These rings contain information about which atoms are bonded together chemically.

Biochemists use graphical sketches in their everyday work, using perspective views and wedges to give a three-dimensional effect. Up to now, only two-dimensional has been used in computer graphics for biochemists' use as a general rule, although three-dimensional software does exist especially for modelling and pattern matching in drug preparation. One of the reasons for the predominance of only two-dimensional is that experimental three-dimensional features are not always known. Another reason is that most commercial databases represent structures using connection tables which generate only two-dimensional pictures. Adequate techniques do not yet exist for the effective representation and storage of three-dimensional structures at bearable costs. Recently the pressure of demand for three-dimensional graphics has increased. This has been caused by the fact that three-

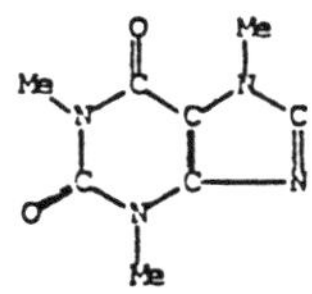

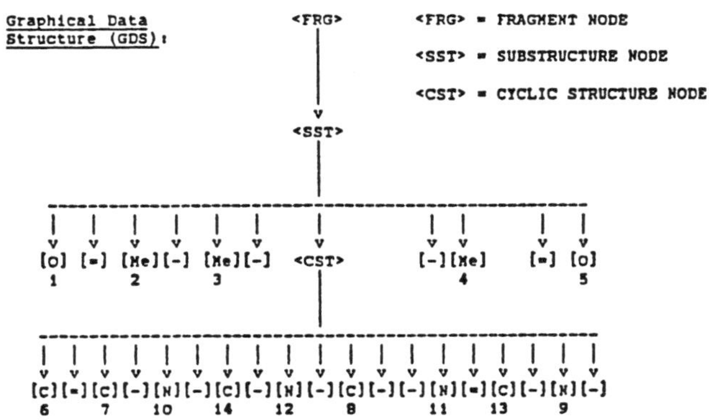

Connection Table:

NOD	SYM	******CONNECTIONS****** NOD/BON	NOD/BON	NOD/BON
1	O	8 CDE		
2	C	9 CSE		
3	C	10 CSE		
4	C	12 CSE		
5	O	14 CDE		
6	C	8 RSE	9 RSE	7 RDE
7	C	10 RSE	11 RSE	6 RDE
8	C	1 CDE	12 RSE	6 RSE
9	N	2 CSE	13 RSE	6 RSE
10	N	3 CSE	14 RSE	7 RSE
11	N	13 RDE	7 RSE	
12	N	4 CSE	14 RSE	8 RSE
13	C	11 RDE	9 RSE	
14	C	5 CDE	12 RSE	10 RSE

Figure 4.7. Graphical Data Structure (GDS) and connection table for caffeine. (Source: Haines, 1989, p.132).

dimensional structure is a necessary ingredient for all theoretical calculations such as molecular mechanics and quantum mechanics to derive optimal geometries (Kao et al., 1985, p.132). It is now possible to create theoretical structures which are as good as experimentally derived ones. So three-dimensional molecular structures are necessary for detailed molecular modelling. Indeed, this seems to be the current situation: because of the huge size of databases three-dimensional storage would prove too costly so only two-dimensional structures are stored. However, for modelling purposes (i.e. manipulating only a few structures at a time, which uses up little computer memory) three-dimensional can be used. Exceptionally, for relatively small databases such as the new carbohydrate databank which will have 6,000 structures compared with the nine or ten million structures in CAS Online, storage of three-dimensional structures may be feasible.

One way to get three-dimensional structures is to derive them from two-dimensional information. Unfortunately while it is possible to go from three-dimensional to two-dimensional the reverse process is not fail-safe. So as a result research has concentrated on programs which make it easier to build three-dimensional structures. It is probable that the next step in biochemical software will involve advances in either the representation or the construction of three-dimensional structures (Kao et al., 1985).

In the past much of the comprehensive three-dimensional structural information has come from X-ray crystallography. In biological molecules the important thing has been the sequence of nucleotide bases in a nucleic acid or of the amino acid residues in a protein or the chemical structure of groups contained within the protein (but not covalently linked to it), together with knowledge of molecular shape, how molecules fit together, and how chemical reactivity and catalysis are controlled.

However three-dimensional information is now being derived also from Raman spectroscopy, lasers (Spiro, 1989) and nuclear magnetic resonance. The latter source is now established on its own database and is becoming widely used for structure identification. Raman spectroscopy can monitor key structural features such as conformation and strength of macromolecular interactions. Spiro comments that the scope of laser applications in biology will expand significantly in the near future. These new sources of information will increase the detail at which three-dimensional structures can be represented and will undoubtedly increase the representation problem discussed above, making worse the computational and (most importantly) memory limitations involved when databases handle three-dimensional structures.

An example of a system which provides for three dimensional representation is Chem-X (Molecular Design Ltd,Unit 12, 7 West Way, Oxford OX2 0JB). Chem-X is a major molecular modelling system used by many large chemical companies and research centres. The system can under certain conditions create three dimensional models of chemical structures from two dimensional structural data. Data on atom types and connectivity are usually read from existing two dimensional structure databases such as MACCS-II, OSAC or DARC. Data can be entered in SMILES notation also. The software has several search and modelling programs. Among these are ChemDBS-3D for 3-D database search, AMBER for molecular mechanics and dynamics studies of macromolecules, and ChemProtein for protein modelling.

It is worth emphasising that three-dimensional information is still quite rare.The data are available from a limited number of experimental techniques, and for a small proportion of chemical compounds. The best application of three-dimensional data is in molecular modelling. Theoretical information about torsion strengths in three-dimensional structures, the kinetic properties of enzymes, and the like is used to make predictions about properties and spatial arrangements. For example Uhlenbeck et al., (1983) working with the bacteriophage MS2 variant R17 have shown that all the coat binding activity is contained in just 19 bases which can be stably folded into a single stem-loop structure. The fragment is bound in a sequence-specific way and interaction involving the coat protein can be observed using filter-binding assays. This system is one of the best understood sequence-specific RNA-protein systems. Uhlenbeck et al., have suggested a model to characterise some of the sequence properties of this system, so that it is suitable for three-dimensional predictions. Haneef et al., (1989) have used three-dimensional modelling techniques to predict several features of Uhlenbeck's model with some success and claim it can be generalised to loop regions in proteins and nucleic acids. Another example of three-dimensional modelling is that by Ferenczy and Morris (1989) who constructed a model of the active site of cytochrome P-450 nifedipine oxidase (P-450 NF). This enzyme plays an important part in the detoxification of drugs used in the treatment of hypertension and angina pectoris. Its sequence showed homology with a bacterial cytochrome P-450 (from Pseudomonas putida) called cytochrome P-450 camphor 5'-exo-hydroxylase (P450 CAM) for which a crystal structure has been reported. This study is based on the assumption that the catalytic cycle does not depend on the source of the enzyme, so that the structure of the active site of mammalian P-450 can be modelled using a bacterial P-450 with known three-dimensional structure. The substrates were docked into the binding pocket and the forces between the substrates and the enzyme were analysed by means of molecular mechanical energy minimisation. Several substrates, including nifedipine, testosterone, quinidine, and Sbenzphetamine, were compared for their degree of fit in the P-450 (NF) active site and the results analysed for their theoretical implications.

Most three-dimensional data derived from diffraction results are contained in the main structural databases. Although created mainly as stores for three-dimensional coordinate data, these crystallographic databases all contain bibliographic and chemical information. The Cambridge Structural Database (CSD) is the largest single source of three-dimensional data (74,000 entries), and is used extensively for modelling purposes. The information includes two-dimensional data (connection table and chemical structure diagram) and three-dimensional data (crystallographic coordinates and symmetry, etc).

Three-dimensional data include data on mean geometry. This is important because it gives much molecular information about bonding and substituent effects. In crystallography it is a means by which new structural results are assessed. Mean geometry has been used, for example, to make new tables of mean bond lengths in organics, organometallics and metal complexes. Also mean geometry is useful in modelling. Within the molecule, it provides data on which three-dimensional models of unknown molecules may be constructed from relevant subfragments, prior to some energy minimisation process. At the intermolecular level, mean geometry derived from crystallographic studies provides the only experimental data by which the geometrical characteristics of hydrogen-bonded and non-bonded systems can be studied systematically (Allen and Lynch, 1989).

4.2.2. Screen representation of graphics

The demand for graphical screen representation of results is increasing. For example chemical patent databases are now provided with graphics. In the Derwent World Patent Index, graphical information has been stored since 1987. Search of graphical features has been created for the Markush - Darc system which has been developed jointly by Telesystemes-Questel and the French Patent Office (INPI).

Graphical screen representation of molecular information is available in both two-dimensional form and three-dimensional form. An example of two-dimensional representation is the Chemical Structure Drawing Interface (CSDI) developed by Hampden Data Services and a possible contender as the standard user interface of the future. A standard international format for data exchange, despite a market-place notable for its variety of formats, has been agreed between ICI, Fisons, Glaxo, CAS, Derwent, Molecular Design and Hampden Data Services. It is in the Standard Molecular Data (SMD) format (see for example Town, 1989, p.1118). Three-dimensional representations are used for CPK surface representations, van der Waals dot surface representations, and van der Waals representation of specific atoms within stick structures (see for example Town, 1989, p.1119).

Three-dimensional representation on screen is still at a very early stage of development. Software is in the stage of adapting to the move away from cheap PCs with medium resolution graphics to workstations with high resolution graphics. One type of screen representation problem is the representation of similarities between two molecules, and of docking processes. For example, a standard problem in drug design is the modelling of three-dimensional interactions between specific small molecules with biologically-active macromolecules that have important roles in disease states, in particular interactions such as three-dimensional binding to a receptor (Hassall, 1985). One solution to the problem of representing molecular movement is to display two structures separately and then together as an overlay to demonstate docking processes and to indicate how natural substrates and inhibitors interact (see for example Vinter and Harris, 1989, p.1111).

Another problem is that of the screen representation of molecular movement. Molecules move, vibrate, interconvert, tautomerise, and internally rotate, on a timescale in keeping with many other molecular processes such as docking, enzymatic reactions caused by docking, and receptor and catalyst complexations. One answer to this problem is to display an overlay of transient structures (Vinter and Harris, 1989, p.1113). The picture shows the breathing concertina motion evocative of movement. Compressed and stretched structures can be distinguished by the use of different colours on screen.

The development of workstations with several processors working in parallel and with high resolution graphics and with sophisticated Unix operating systems means that three-dimensional colour graphics will develop rapidly for the representation of molecular information.

4.3. INTERCONVERSION OF STRUCTURE REPRESENTATIONS

Ambiguous representation, and the unambiguous use of connection tables or linear notations, have advantages and disadvantages that recommend their use in particular circumstances. For example, a linear notation such as WLN is compact and unambiguous, but does not allow flexible search techniques possible with connection tables. Classes of compounds, particularly those described by Markush formulae in patents, are best described by means of fragmentation codes. While connection tables have some advantages, they are not economical, a disadvantage of all representations providing graphical information.

For this reason it is best to provide interconversion algorithms in any system, to move from one representation to another as the user's purpose changes. An instance of this is provided by the CROSSBOW system. In CROSSBOW, WLN was used for structure input and registration, but was converted into CROSSBOW connection tables for structure searching and display. Other arguments for interconversion exist also. For example the user may wish to make comparisons between an external and an in-house database, where the forms of representation differ. Or the user may wish to copy a file from one database to another, again where the representations differ.

In terms of automatic conversion from a connection table to WLN the success rate is only 90 percent. This is mainly because of the fact that WLN uses context-sensitive ring descriptions which are only implicit in connection tables. For the same sorts of reasons automatic conversion from connection table into systematic name is difficult. However, the reverse conversion, from CA Index Names to CAS connection tables is a routine operation (Vander Stouw et al., 1974).

4.4. REGISTRATION

Registration is the process of giving each distinct chemical structure a unique identifying name or number. Information which is being input to the database can then be scanned to see if the chemical structure is a new one. If it is new, an identifier is given to it and it is added to the database. Otherwise the information is assigned to the existing identifier.

In large databases the registration process is required to be accurate and complete. However many smaller databases do not meet this ideal state, leading to wasteful or unsuccessful searches.

One method of registration is to use an alphabetical name index and presumes a unique or canonical structural representation. The alternative method is similar to the creation of a molecular formula index. It is known as the isomer sort technique. It does not depend upon a unique representation. Rather the determination of whether or not the structure is new to the file is carried out by atom-by-atom search. The search method is easilly amenable to parallelisation. The main commercial online database hosts such as DIALOG and STN have banks of large and powerful processors. When a user chooses a domain file from which to search, the file size determines how many processors are assigned simultaneously to the task. Search then proceeds by atom-by-atom search. Each host processor deals with a different atom in the structure. In this way search times are not correlated with the size of the file to be searched. If this desirable state of affairs did not exist, users would be put off using very large databases, and in the desire to save time, would accept second best with smaller databases, a situation unacceptable to database hosts and producers, who wish to make very large databases as attractive as possible.

Structures in molecular formulae are distributed unevenly which is a disadvantage for the isomer sort technique. Many chemical formulae are unique, representing a single chemical compound. But this only occurs in the case of structures of elements which occur infrequently or which occur in infrequent combinations. The number of isomers in molecular formula groups may reach a thousand in very large files, so that comparison with the candidate structure becomes impractical. Oppenheim (1979, p.384) provides an example of a molecular formula which represents many different compounds. So other characteristics have to be used to divide up these large groups and make each subgroup searchable.

4.4.1. Registration with canonical description

With this method the linear notation and connection table representation used are assumed to be unique for each distinct structure. Also each atom within the structure is assigned a unique (though of course arbitrary) number. The file is then sorted alphabetically using the linear notation name as the key identifier. Many algorithms for this process exist. For example the Chemical Abstracts Service (CAS) Compound Registry Service uses the following method (Schulz, 1988):

(1) All non-hydrogen atoms in the molecule are first numbered serially.

(2) One atom is selected, given number 1, then ones connected to atom 1 are selected and given number 2 and so on.

(3) The unnumbered atoms connected to atom 2 are numbered and the process repeated till all atoms have numbers.

(4) Where a structure is discontinuous, all within-group atoms are given consecutive numbers.

4.4.2. Isomer sort registration

The efficiency of this technique depends on how large subgroups are subdivided. The more evenly compounds are distributed among the subgroups the more efficient will be sorting using this method.

To do this two alternative methods are adopted. The first method is to use uniformly more information about each structure than that contained in the molecular formula. The other approach is to adjust the level of detailed description to the expected size of a subgroup. This is more economical and is justified by the fact that many subgroups contain only one structure. Fortunately the size of a subgroup is determined by its chemical characteristics. The approach involves sampling each file for its chemical characteristics. The most highly populated molecular formula groups are those of carbon and hydrogen, together with either oxygen or nitrogen, or both.

4.4.3. Translation between systematic nomenclatures

As has already been described, there are several alternative methods of machine representation of molecular information. There are the nomenclatures used for formulae, naming compounds and linear notations such as WLN or IUPAC. Molecular structures are also represented by means of connection tables, and in fact these tables are the most widely used method of representation of topological information. The major abstracting services however do not have a completely systematic method of inter- conversion between different structural representations required. While they use automated methods these methods frequently default to ad hoc algorithms using lookup tables of morphemes (Cooke-Fox et al., 1989). For example this is the current position of translation procedures at CAS for converting systematic names to atom-bond connection tables. Routines are used to edit and verify newly appearing index names. The CAS procedures are based on lookup tables of basic word roots and once these are found their semantic meaning is identified by ad hoc methods. The trouble with this approach is that special routines have to be used frequently, for example when bracketed, highly branched constituents are encountered). As a concluding indicator of the problems of interconversion between nomenclatures, a recent evaluation (Cooke-Fox et al., 1989 p.102) is that existing programs are large, complicated, and not completely reliable.

4.4.4. Automatic Indexing

The foregoing problem of creating a completely unique representation schema, especially the problem that commonly encountered substances occupy large subgroups, has a significant effect upon the problem of automatic indexing. Although the problem is small for some substances such as pure isomers, chemical nomenclature, and interconversion schemes between alternative nomenclatures, become considerably less rigorous for commonly found substances. The fact that human error is so common means that it is very difficult to make unambiguous rules which can be automated. Buntrock (1989) has for example noted that although the chemistry of C4 compounds is more ordinary than for the "stranger" C1 compounds, the nomenclature problems are much greater, owing to the abundance and commonness of C4 compounds. He documented indexing errors for butanols, and found that indexing and structure representation were inconsistent for the butenes, and that several abstracting services made persistent errors for butene monomers and polymers ("polybutenes"). He commented:

> "Both n-butanol and isobutanol are primary alcohols. However an error of perception is often made upon examination of the structure of secondary butanol (2-butanol). The branched chain is noted, and the compound is misattributed "isobutanol". Such errors can even creep into abstracting and indexing services. For example,in the past 20 years of Chemical Abstracts, more than 1700 documents have been indexed to sec-butanol and not to isobutanol. (Unless stated otherwise, use of Chemical Abstracts Registry Numbers (CASRN) was the search strategy of choice because of specificity of indexing. All searches were performed on the CAS Online files as loaded on the STN network). However eight of those references are also indexed with the term "isobutanol". All but one are apparently misattributed to isobutanol because 2-butanol is indeed described. The remaining reference describes separation of isobutylene, isobutanol, and tertiary butanol, and the Registry Number for 2-butanol is apparently misassigned" (Buntrock, 1989, p.72).

These errors are relatively infrequent, but in the butene polymers

> "confusion reigns, largely due to inadequate, inconsistent or poorly described nomenclature used in the original publications" (Buntrock, 1989, p.73).

In conclusion then it is important to realise that molecular databases suffer from a number of problems. For example, when the user switches from one search key to another, necessitating interconversion to be done between internal representations, the algorithm used is not completely fail-safe. Even on the primary key many databases have ambiguities resulting in gaps in results or completely wrong answers. Moreover, automated indexing still has a number of deficiencies leading to incorrect assignment. While these problems crop up in only a very low percentage of all search results, it is important to remember that the utility of a database to end users depends not only on its size but also upon the quality of information stored and reliably retrieved.

4.5. REFERENCES

Allen F.H., and Lynch M.F., 1989, "The storage and retrieval of chemical structures", Chemistry in Britain, 11,1101-1108.

Ash J.E., and Hyde E., (Eds), 1975, Chemical information systems, Ellis Horwood, Chichester.

Ash J.E., Chubb P.A., Ward S.E., Welford S.M., Willett P., 1985, Communications, Storage, and Retrieval of Chemical Information, Ellis Horwood, Chichester.

Buntrock R.E., 1989, "Documentation and Indexing of C4 compounds : pathways and pitfalls", J.Chem.Inf.Comput.Sci., 29, 72-78.

Cooke-Fox D.I., Kirby G.H., Rayner J.D., 1989, "Computer translation of IUPAC systematic organic chemical nomenclatures. 1.Introduction and background to a grammar-based approach", J.Chem.Inf.Comput.Sci., 29, 101-105.

Dittmar P.G., Mockus J., and Couvreur K.M., 1977, "An algorithmic computer graphics program for generating chemical structure diagrams", J.Chem.Inf.Comput.Sci., 17, 186-192.

Ferenczy G.G., and Morris G.M., 1989, "The active site of cytochrome P-450 nifedipine oxidase: a model building study", J.Mol.Graphics, 7, 206-211.

Gleschen D.P., 1989, "Enhancing treatment of nucleic acids and proteins in the CAS Registry File", Chemical Information Bulletin, April, 22.

Haines R.C., 1989, "Computer graphics for processing chemical substance information", Computers Chem., 13, 129-139.

Haneef I.,Talbot S.J., Stockley P.G., 1989, "Modelling loop structures in proteins and nucleic acids: an RNA stem-loop", J.Mol.Graphics, 7, 186-195.

Hassall C.H., 1985, "Computer graphics as an aid to drug design", Chemistry in Britain, January, 39-44.

Kao J., Day V., and Watt L., 1985, "Experience in developing an in-house molecular information and modelling system", J. Chem. Inf. Comput. Sci., 25, 129-135.

Lynch M.F., Harrison J.M., Town W.G., and Ash J.E., 1971, Computer handling of chemical structure information, MacDonald, London.

Mockus J., Isenburg A.C.,and Vander Stouw G.G., 1981, "Algorithmic Generation of Chemical Abstracts Index Names: 1.General Design", J.Chem.Inf.Comput.Sci., 21, 183-195.

Oppenheim C., 1979, "Methods for chemical substance searching online: 1: The basics", Online Review, 3, 4, 381-387.

Rossler S., and Kolb A., 1970, "The GREMAS System, an integral part of the IDC System for Chemical Documentation", J.Chem.Doc., 10, 128-134.

Rowlett R.J., 1975, "CA (Chemical Abstracts) Nomenclature", Chem.Eng.News, 53, (3), 46-47.

Schulz H., 1988, From CA to CAS Online, VCH Verlagsgesellschaft.

Spiro T.G., 1989, "Probing biological molecules with lasers", Chemistry in Britain, 25, 602-605.

Town W.G., 1989, "Microcomputers and information systems", Chemistry in Britain, November, 1118-1121.

Uhlenbeck O.C.,Carey J.,Romaniuk P.J.,Lowary P.T., and Beckett D.B., 1983, J.Biomol.Struct.Dyn., 1, 539.

Vander Stouw G.G., Elliott P.M., and Isenberg A.C., 1974, "Automated conversion of chemical substances names to atom-bond connection tables", J.Chem.Doc., 14, 185-193.

Vander Stouw G.G., Gustafson C., Rule J.D., and Watson C.E., 1976, "The CAS chemical registry system IV. Use of the registry system to support the preparation of index nomenclature", J.Chem.Inf.Comput.Sci., 16, 213-218.

Vinter J.G., and Harris M., 1989, "The graphical display of chemical structures", Chemistry in Britain, November, 1111-1117.

Weininger D., 1988, "SMILES, a chemical language and information system. 1, Introduction to methodology and encoding rules", _J.Chem.Inf.Comput.Sci._, 28, 31-36.

Weininger D., Weininger A., and Weininger J.L., 1989, "SMILES. 2, Algorithm for generation of unique SMILES notation", _J.Chem.Inf.Comput.Sci._, 29, 97-101.

Chapter 5:

Database Searching in Biochemistry and Molecular Science

5.1. BIBLIOGRAPHIC SEARCHING

Only bibliographic databases give end users the complete freedom they often require to search for any key word or structure alternatively or in combination.

An example given by Chambers (1984) involved a search using

"Pharoah's ants" and
"C20 H32" and
"ring size - 6,10, or 14" and
"pheromone".

He wanted to identify the structure of the chemical which had been isolated from Pharoah's ants and which was thought to be pheromone (a substance thought likely to alter their behaviour).

Chambers found that manual, hard-copy search would have involved searching 116 different structures (by means of reading their indexed journal articles) since all had molecular formula C20 H32. An online search meant searching only 13 structures, because ring size was more specific as an index term. A limitation of online search was the lack of an index for "diterpene" (it was known that the substance was a monocyclic diterpene). Another problem of online search is the alternative names for the same chemical (e.g. hexanoic acid, caproic acid). Also CAS Online gives different registry numbers to different stereoisomers of the same compound, a problem of online retrieval touched on earlier. Several studies (for example Borkent et al., 1988) have reported that the degree of overlap between databases is surprisingly small, necessitating searching of several databases to ensure comprehensive coverage. This is a big disadvantage of online searching which pushes search times up and which discourages many scientists from doing fully comprehensive searching. For this reason several database hosts (for example DIALOG) have introduced online multifile capability.

Bawden and Brock (1985) did a comparative study of online databases using eight different search queries. For example, one query was

> "mammalian toxicity of orally administered cinnamic acids, salts or alkyl esters"

Their conclusion was that

> "(1) Online searching is an essential component of virtually all chemical toxicology searches but must be complemented by printed sources,
> (2) a multifile approach is virtually always required,
> (3) the specialised searching facilities of each database should be utilised,
> (4) profiles of search terms are necessary for many concepts,
> (5) databanks have significant advantages over bibliographic databases for many searches". (Bawden and Brock, 1985, p.35).

The last point is important in that many database vendors are putting up files which have mixtures of full text, reference citations, and topographic structure information. As Bawden and Brock (1985) found

> "Databanks, with their record organisation by chemical structure, specialised search facilities, and direct link of structure to toxicity data, were particularly valuable for some queries" (Bawden and Brock, 1985, p.35).

Another particularly valuable property of online searching is that of browsing, ie looking for something which is partly predefined by the user and partly novel, if possible. This is a field of active current research. Several algorithms exist which look not for perfect matching of candidate structures and/or text strings, but which seek targets that have the "nearest" characteristics. The approach taken is to exclude targets which perfectly match the input candidate, thus getting all "neighbours" to the described target. An example of such a nearest-neighbour search facility is one which calculates a measure of inter-molecular structural similarity between the query compound and the molecules in the database which is being searched (Willett et al., 1986). The major problem is that search time may be excessively long, so that a preliminary substructure search is necessary first.

Because biochemical information comprises structure and function it is natural to focus on this relationship in the construction of production rules when developing artificial intelligence applications. For example, one production rule might be

if SUBSTANCE is used for USE
in some FIELD because of some PROPERTY then
....

Smith and Chignell (1984, p.101) claim that

> "The goal is to develop a computer program that can read a database directly and organise the available knowledge into sketchy scripts ....that can then be used for the development of a search system. At first, this approach will be applied to the controlled-vocabulary terms and associated free text modifications. It is hypothesised that the grammar used by indexers is sufficiently constrained to permit the development of a computer program that can read and "understand" the contents....Simple realisations of such a deep structure might include
> PYROLIZIDINE ALKALOIDS are used as NEOPLASM INHIBITORS
> GLYCOALKALOIDS are used as PESTICIDES in AGRICULTURE"

This claim is startling because current advanced expert systems such as the medical expert system MYCIN (Shortliffe, 1976) and the chemical structure identification system DENDRAL (Lindsay _et_ _al_., 1980) use production rules that have been created by experts. Library science appears to be the area where query text strings are sufficiently limited in style that any expert system would not be held back by the current poor state of the art knowledge of how to deal with natural language processing. It is the area where a fairly new type of expert system is about to be developed - one which can write the production rules itself by reading through a database's indexes. In 1988 an initial prototype for such an expert system, called ID3, was developed (Schlieper and Isenhour, 1988). Their algorithm automatically creates production rules and can distinguish when attributes and their values do not sufficiently span a given domain. It seems that more powerful and sophisticated programs will undoubtedly follow.

However, even without expert databases there are many areas where even simple online databases represent an improvement over existing arrangements. One current project at Bath University (supported by the U.K. Computer Board) is the online version of the _Science Citation Index_, using a database of scientific references compiled by the Institute for Scientific Information (ISI) in Philadelphia, USA. Currently ISI subscribers who phone ISI must ask librarians there to search the database in order to minimise telephone charges. Once the project is completed there will be desktop access for several hundred concurrent users. Initially access will be via JANET and will be free to academic users.

5.2. PATENT SEARCHING

Patents in molecular science usually refer to large classes of molecules, all of which are claimed to have some particular property. The wide scope is necessary to prevent competitors "getting round" a patent by marketing a product similar to that patented. Because of this many patents lay claim to millions of individual substances for a particular property. This causes great problems for data retrieval in patent searches. Figure 5.1 shows a typical generic structure description of a patent.

Such structures are called "Markush" structures. There is a constant part (group of atoms and bonds common to all compounds in the class) and some variable parts called R1, R2, and so on. A set of alternative values is then given to these variable parts. They may be variable in their chemical constituents, in their position of attachment to the constant part, and in the number of times they occur.

In the past, storage and retrieval of Markush structures in patents were based on manually prepared fragmentation codes, since automated

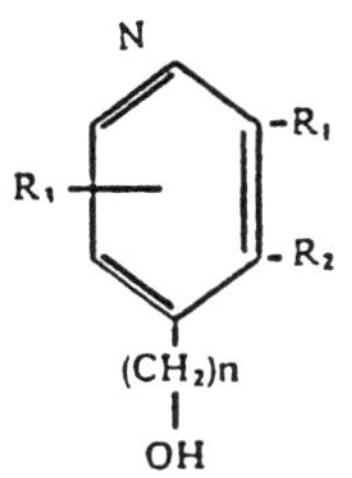

R_1 is H, OMe, $OCH_2CH_2NH_2$, or 2-8C alkyl, preferably ethyl; R_2 is amino, heterocyclyl, phenyl optionally substituted by halo, or R_1 and R_2 together is alkylene; R_3 is 1 or 2 chloro or 1 or 2 bromo; n is 1-4; with the proviso that if R_1 is hydrogen then R_2 is amino.

Figure 5.1. A Markush Structure Description. (Source: Barnard, 1987, p.32).
Reprinted with permission from Database, Vol. 10, Barnard J.M., "Online graphical searching of Markush structures in patents", Copyright (c) 1987, Online Inc.

linear notations such as Wiswesser Line Notation (WLN) were generally inapplicable. These systems have therefore suffered from the general disadvantages of fragmentation codes: ambiguous representation, inconvenience to the user in having to get acquainted with a complicated fragmentation code, and non-portability from database to database (see Chapter 5). There was also the disadvantage that such systems (except GREMAS) could not distinguish between the constant and the variable part.

However in 1989 a Markush structure database was made available by STN International. It allows topological (i.e. graphical) storage and retrieval. This database will complement the only other existing such facility, provided by Derwent/Telesystemes/INPI since 1987.

The STN database uses GENSAL, a completely automated approach to Markush structure file input and output. GENSAL, developed by Barnard, Lynch and others at Sheffield, represents Markush structures in a form close to the original patent specification. Although INPI at present uses the manually input fragmentation codes from GREMAS, they too are investigating using GENSAL. However, alternative approaches are still under active investigation. For example, Tokizane et al., (1987) use a method which represents generic topological information containing fixed or variable parts in the form of bit patterns allowing rapid search through these compressed data structures. Lynch's group has also been working on parallelisation of patent searching and this must be looked on as one of the promising areas for the future in terms of bringing down search times from the average 5 minutes at present. Parallelisation is the breaking of the search algorithm down into pieces which can run in parallel thus making total run time shorter. Another approach to search time minimisation by the same group involves summarising whole structure information in the form of "reduced chemical graphs". The gross structural features are summarised, usually in terms of only the ring or the non-ring components, giving a tree structure in which each mode is either a cyclic or acyclic component. Other approaches have been made to graph reduction by others, including those in which the components are formed from the separate aggregates of carbon atoms and of heteroatoms. Reduced chemical graphs are essentially multi-graphs, in which the variables are similarly summarised.

Online patent searching has been found to be amenable to applications of artificial intelligence. Ardis (1990) has developed a working expert system which has been in continuous use since 1987. The program carries a number of functions:

(1) There is an initial test for patentability, considering the information which has been supplied by the user.

(2) The system determines how sophisticated the user is, and responds accordingly.

(3) The user supplies his main search goals and what constitutes the search problem is automatically deduced by the system.

(4) There is a search and data retrieval.

The program acts as an automatic gateway because it performs the following functions without the interference of the user:

(1) Given a series of requirements or rules, the appropriate database is automatically chosen. There is now a considerable movement within the software (and computer hardware) community towards 'open systems', where the supplier's copyright (and unique features) are waived (and standardised) in favour of making such software (and hardware) more widely useable and therefore more valuable.

(2) Given the database and type of search chosen, the system determines the search commands necessary.

(3) The telecommunications link-up and log on are made automatically.

(4) the user provides the parameters for outputting the information.

(5) The system then runs the search and outputs the results, logs off, and does any remaining chores such as printing, downloading data to PC, and so on.

Ardis (1990) is considering extending the system for dealing not only with patents but also with trademarks, and for front-ending the system to the new CD-ROM product developed by the U.S. Patent and Trademark Office.

5.3. SUBSTRUCTURE SEARCHING

The aim of substructure searching is to identify all compounds in a collection which contain a given partial structure. Figure 5.2 shows an example of a request for all structures containing a p-benzoquinone ring and a phenyl group substituted with fluorine or chlorine in the para-position. It also shows the use of and/or/not (Boolean) logic.

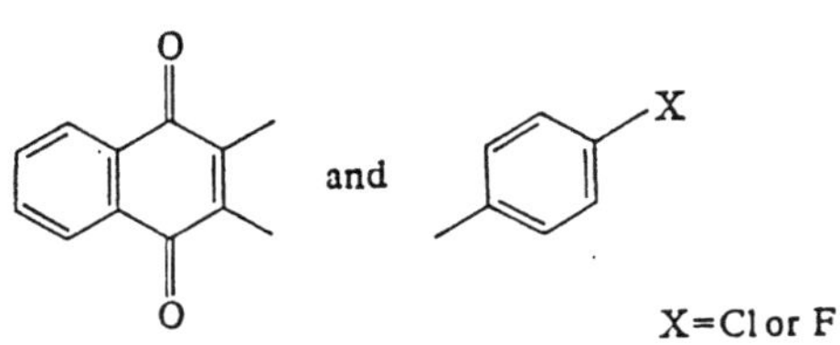

Figure 5.2. Example of substructure query.
(Source: Lynch et al., 1971, p.67).
Reprinted with permission from Lynch M.F., Harrison J.M., Town W.G., and Ash J.E., "Computer handling of chemical structure information", Copyright (c) 1971, Macdonald.

Both linear notations and connection tables have been extensively used in substructure searching. So it is as well to remember from the previous chapter the problems of creating unambiguous descriptions of molecular topology. These problems render substructure searches only incompletely effective.

It has been argued that these problems will only go away when a general standardised method has been created for classifying substructures. That is to say the need is for more theory, not more programs. Dubois et al., (1987, p.74) suggest

> "The notion of substructure remains complex and difficult to define since it covers very diverse aspects: joined or disjoined substructures, completely defined or fuzzy substructures, two or three dimensional substructures. A substructure can be considered in the structural space alone or in its association with certain physical characteristics such as pK, spectroscopic data, partition coefficients, or pharmacodynamic activity. At that point, certain properties associated with local functions of molecular structure are expressed. An overall taxonomy, bearing on the whole set of substructures, would be extremely useful, since substructures are essential factors in the basic software of both documentation and artificial intelligence."

However, the aims of using a substructure relating to spectroscopy data is different from the aims of a substructure used in computer-aided synthesis. The pressure of diverse end user needs has led to ad hoc creation of incompatible substructure schemas, resulting in very simple and similar names taking on more than one meaning. This has generated pressure for a general unifying conceptual framework. One important piece in this theoretical jigsaw has been the concept of Fragments Reduced to an Environment that is Limited (FRELs). These are substructures which are defined concentrically around a focus. The focus is usually an atom or two bonded atoms. Other foci are possible (e.g. rings). Thus the FREL consists of a focus and an environment. The FREL concept has been implemented in a piece of software called Description Acquisition Retrieval and Conception (DARC). It seems likely that further theoretical advances are required before substructure searching can be improved. The DARC system gives online access to the EUROCAS chemical structures database based on the CAS Registry File. DARC allows chemists to perform substructure searches with text or graphic input of structures. Screens are generated which will find all structures which possess the requested substructure(s). Searching is then done by atom-by-atom or bond-by-bond search, resulting in a list of CAS Registry Numbers.

Unfortunately, substructure search of very large databases is too costly to be practicable. This means that the database must be reduced in some way, using a screening process, i.e. a rapidly executable preliminary search which finds a correspondence between simple characteristics of the query and each structure on the file so as to reduce to a minimum the number of candidates on which a full search is performed.

5.3.1. Screens

The problem of making useful screens is the wide variety in frequency of occurrence of the different elements in chemical structures. Carbon, oxygen, and nitrogen account for 95% of the total number of atoms. So a search for a structure containing a rare element like iodine would be highly selective (but also, unfortunately extremely unlikely that such a screen would be needed) compared with a screen for structures containing carbon, oxygen, or nitrigen (Lynch _et al_., 1971, p.90).

Another problem is that the molecular formula is also a weak screen. A structure must contain a relatively infrequent element or a large number of carbon, oxygen, or nitrogen elements before the molecular formula becomes useable. Theoretically, each screen should occur in 50% of all structures in a file. Unfortunately, structural characteristics of files are not random. Also files have special features.

Therefore when we choose sets of screens based on atom groups, size, constituents of rings, and group relationships, these need to bear some statistical relationship with their observed frequency of occurrence within the particular file to be searched. How we choose the screens depends upon the performance of alternative screening methods which have been tested in realistic circumstances. But, because the file characteristics make their own individual impression upon the screen's effectiveness, it becomes necessary to use statistical methods such as discriminant analysis, factor analysis, and cluster analysis. Given a large number of items scored on a large number of attributes discriminant analysis will tell us which attributes are most associated with a particular attribute. Factor analysis puts the attributes into bundles of similar attributes, and cluster analysis goes one step further and shows the family resemblances of these attribute bundles by drawing a family tree relating the bundles together in terms of similarity and dissimilarity. Use of these techniques allows the search algorithm to be applied to any file. Any file can thus be quickly scanned and the items within it subjected to one of the types of statistical analysis described. Research reported by Willett et al., (1986) claims significantly improved performance using cluster analysis as a preliminary screening aid.

5.3.2. Structure elucidation

The aim of structure elucidation is to make easier the process by which the biochemist and molecular scientist deduce the structure of an unknown substance from spectroscopic and chemical data. Central to the process are programs which generate all molecular structures which satisfy the constraints that are contained in our state of knowledge of the unknown structure. These programs are called structure generators. They generate all plausible candidates (Funutsu et al., 1988, 1989). The problem then becomes that of narrowing down the candidate set to one chosen structure. When this is not possible the biochemist must plan new experiments to gain new data about his unknown substance. However, one of the problems is that often the number of candidate structures is very large, providing few clues. For this reason, programs exist to classify candidate sets into subsets satisfying certain features, selected by the user. Alternatively, the choice of these features can be made by the program. This latter approach requires the program to place the candidate structures into disjoint sets satisfied by as many features as possible (Lipkus and Munk, 1988).

5.3.3. Pattern matching searches

Because so much of molecular and biochemical information is graphical, either in two or three dimensions, and because the end user demands the facility of carrying out structural modifications during search, the problem naturally lends itself to pattern matching algorithms. One of the applications of these methods comes in studies of reaction pathways where biochemical substances change from one reaction to the next in a series, or pathway. Stimulation or inhibition of reactions is often a principal aim of biochemical experiments.

For example, Cody (1986) studied the binding of antifolates in the active sites of the dihydrofolate reductase (DHFR) of chicken liver. DHFR is an enzyme which is necessary for cell growth. He found which substances occupied hydrophobic (water avoiding) pockets (Hulme, 1990, p.278). Using pattern matching on an online database he was able to explore the size and position of the protein environment near these pockets. He concluded that relevant chemical substances failed to make optimal contacts with the functional groups surrounding the pocket. He then suggested how to design alternative antifolates which could make specific optimal contacts in this part of the protein. For example he compared the naturally occurring chicken liver substrate folic acid and the anticancer chemotherapeutic inhibitor methotrexate. They have very similar chemical diagrams. But they have very different reactions to particular drugs. Also their three dimensional structures are different, which shows the superiority of three dimensional over two dimensional information, a point already made in Chapter 4. Cody (1986, p.69) suggests

> "Marked differences in the inhibitory potency have been observed concomitant with small changes in the chemical structures of these antifolate drugs".

In the past, structure-activity relationships were modelled by means of multivariate statistical techniques such as multiple regression or discriminant analysis. However, these methods made little use of stereochemical and spatial information. For this reason current work has focussed on making use of the now available molecular graphics capabilities of modelling and search programs. These provide more insight than statistical methods into the 3D interactions between active molecules and biological receptor sites. One application of molecular graphics is in analysing and displaying pharmacores, the set of structural features in the drug molecule that is recognised at the receptor site. Jakes and Willett (1986, p.12) explain

> "The most widely used method of pharmaphoric pattern identification is to compare the energetically feasible conformations of these molecules which are known to exhibit the requisite activity so as to obtain a map of the receptor site".

Their paper describes a method of developing a 3D substructure search system to retrieve all those compounds from an online database that contain an input query pharmacophore. This focusses on the selection of inter-atomic distances which are used as descriptors to facilitate the efficient search of online databases. This example emphasises the integrated use of programs in current work. The process includes modelling, then preparing searches, then doing online searches and feeding the results back to the modelling program.

5.3.4. Chemical structure searching and modelling software

CAS Online (STN) and DARC (Telesystemes-Questel) are the largest and most important online structure search systems with graphics input and output. In both the specification of structures of interest can be done by drawing on a graphics terminal or input through a text terminal. The type of terminal used and the generation of query structure are described in detail by Schulz (1988), p.146-162.

STN Express, a PC-based front-end interface for the Messenger software that is used to search STN International databases, was introduced in 1989 (Haines, 1989, p.137). Before the availability of STN Express software, chemical structures were created online with the structure modelling and searching software (Messenger) mounted on STN's mainframe using commands as described in Schulz (1988, p.150). According to Brueggeman (1988) chemical structure queries can now be created locally, in the PC's memory, and then uploaded by STN Express. This new approach is cheaper and more convenient, taking the pressure off the searcher who is naturally worried that his slowness is costing money. Unfortunately, at present

> "Offline-created structures cannot be subsequently modified with STN's STRUCTURE command while online" (Brueggeman, 1988).

Despite this, the facility to prepare search commands before going online is valued by less experienced users. The search commands can be prepared as a batch job while users are waiting for the results of the previous chemical structure search which currently take as long as five minutes. Brueggeman (1988) compared STN Express with the text searching and editing features of alternative packages such as PCPLOT III, STN Communicator (used for graphical searching), and the text-oriented search-aid software like Dialoglink and ProSearch. STN Express only draws structures that can be searched on STN International databases. Molkick (from Springer-Verlag) goes one step further and translates offline-drawn structures into any computer-representation acceptable to any of the three structural query languages (as used by CAS Online, DARC, and ROSDAL) which are used by the five search services (Data-Star, DIALOG, Maxwell ORBIT, Telesystemes-Questel, and STN). Brueggeman (1989) compared Molkick and other packages such as STN Express. One weakness of Molkick is that it is only intended for structure creation, and needs full communication software to go online.

Other software packages relevant to chemistry and biochemistry such as ChemBase, ChemFile, Chemword, HTSS and PSIGen/PSIBase were compared by Heller (1987). ChemBase from Molecular Design Ltd is directed toward the synthetic organic chemist. It offers high performance but is expensive because it requires a colour monitor and graphics board and a laser printer. ChemBase is basically a database manager which instead of allowing the user to manipulate text and numbers

> "allows him to store, manipulate and search the database in terms of complete reactions" (Marshall, 1986).

It can search a database for all entries containing a specified compound, structural unit, functional group or any combination of these. The program also supports more conventional types of database interrogation (e.g. retrieval of all compounds available with molecular weights in a certain range, boiling points, etc) using Boolean operators (see also Rey, 1987). SANDRA (Structure and Reference Analyser) is an IBM-PC package which is designed to help the user look for chemicals in the Beilstein Handbook of Organic Chemistry.

Organic chemists and biochemists are interested in calculating three-dimensional atomic coordinates to build and manipulate molecular models. PC-based packages which allow this are becoming available: Alchemy, by Tripos Associates Inc is one of them. Alchemy enables the user to build molecules from atoms and fragments, modify bonds or by optimisation of molecular structures using an energy minimisation procedure. The program

> "allows the user to change the position of atom centres of molecules either by translation and rotation in three dimensions with the mouse or by superimposition of two molecules" (Turk and Zupan, 1988).

There will be better packages available in the near future. It is reported that CAS is interested in this field and is planning to interface its software and possibly its 9-10 million substance Registry File with similar packages. Haines (1989) writes:

> "a multi-level search algorithm is envisaged, with the final search step being a geometric match of the 3D query structure against the 3D file structure".

5.4. REFERENCES

Ardis S.B., 1990, "Online patent searching: guided by an expert system", Online, March, 56-62.

Barnard J.M., 1987, "Online graphical searching of Markush structures in patents", Database, 10,27-34.

Bawden D., and Brock A.M., 1985, "Chemical toxicology searching: a comparative study of online databases", J. Chem. Inf. Comput. Sci., 25, 31-35.

Borkent J.H., Oukes F., and Noordik J.H., 1988, "Chemical reaction searching compared in REACCS, SYNLIB, and ORAC", J. Chem. Inf. Comput. Sci., 28, 148-150.

Brueggeman P.L., 1988, "STN Express: search-aid software for chemistry", Database Searcher, 4, 15-23.

Brueggeman P., 1989, "Creating chemical structures for online searching with Molkick", Database Searcher, 5, 22-27.

Chambers J., 1984, "A scientist's view of print versus online", Aslib Proceedings, 36, 309-316.

Cody V., 1986, "Computer graphic modelling in drug design: conformational analysis of dihydrodrofolate reductase inhibitors", Journal of Molecular Graphics, 4, 69-73.

Dubois J.-E., Panaye A., and Attias R., 1987, "DARC system: notation of defined and generic substructures; filiation and coding of FREL structure (SS) classes ", J. Chem. Inf. Comput. Sci., 27, 74-82.

Funatsu K., Miyabayashi N., and Sasaki S., 1988, "Further sevelopment of structure generation in the automated structure elucidation system CHEMICS", J.Chem. 28, 18-28.

Funatsu K., Susuta Y., and Sasaki S., 1989, "Introduction of two-dimensional NMR spectral information to an automated structure elucidation system, CHEMICS. Utilization of 2 D-inadequate information", J. Chem. Inf. Comput. Sci., 29,6-11.

Haines R.C., 1989, "Computer graphics for processing chemical substance information", Computers Chem., 13, 129-139.

Heller S.R., 1987, Chemical substructure searching on a PC, 25-32 in Online Information 87, 11th International Online Information Meeting, London 8-10 December 1987, Learned Information.

Hulme E.C., (Ed.), 1990, Receptor biochemistry: a practical approach, I.R.L. Press, Oxford University Press, Oxford.

Jakes S.E., and Willett P., 1986, "Pharmacophoric pattern matching in files of 3D chemical structures: selection of interatomic distance screens", Journal of Molecular Graphics, 4, 12-20.

Lindsay R., Buchanan B.G., Feigenbaum E.A., and Lederberg J., 1980, Applications of artificial intelligence for chemical inference: the DENDRAL project, McGraw-Hill, New York.

Lipkus A.H., and Munk M.E., 1988, "Automated classification of candidate structures for computer-assisted structure elucidation", J. Chem. Inf. Comput. Sci., 28, 9-18.

Lynch M.F., Harrison J.M., Town W.G., and Ash J.E., 1971, Computer handling of chemical structure information, Macdonald, London.

Marshall J.C., 1987, "ChemBase", J. Chem. Inf. Comput. Sci., 27, 47-49.

Rey D., 1987, Applications of personal computer products for chemical data management in the chemist's workstation, 48-61 in Warr W.A., (Ed.), Graphics for chemical structures: integration with text and data, ACS Symposium series 341, American Chemical Society, Washington.

Schlieper W.A., and Isenhour T.L., 1988, "Using analytic data to build expert systems", J. Chem. Inf. Comput. Sci., 28, 159-163.

Schulz H. 1988, From CA to CAS Online, VCH Verlagsgesellschaft.

Shortliffe E.H., 1976, MYCIN: computer-based medical consultations, Elsevier, New York. Based on a PhD thesis, Stanford University, 1974.

Smith P.J., and Chignell M., 1984, "Development of an expert system to aid in searches of the Chemical Abstracts", Proceedings of the 47th ASIS Annual Meeting, 21, 99-102.

Tokizane S.,Monjoh T., Chihara H., 1987, "Computer storage and retrieval of generic chemical structures using structure attributes," J. Chem. Inf. Comput. Sci., 27, 177-187.

Turk D., and Zupan J., 1988, "Alchemy", J. Chem. Inf. Comput. Sci., 28, 116-118.

Willett P., Winterman V., and Bawden D., 1986,"Implementation of nearest-neighbour searching in an online structure search system", J.Chem.Inf.Comput.Sci., 26, 36-41.

Chapter 6:

Using Expert Systems for Database Searching in Molecular Science

6.1. INTRODUCTION

It has been suggested that the use of expert systems will radically alter the way in which information professionals go about their work (Morris and O'Neill,1988). Davies (1989) has raised the intriguing question of whether new knowledge can be created by using novel approaches to information retrieval. Such new knowledge might be discovered by means of information retrieval systems which (i) discover a hidden refutation or qualification of a hypothesis, or which (ii) draw conclusions from an unused piece of evidence, or which (iii) draw conclusions from a series of tests which by themselves are too weak to justify such a conclusion, or which suggest analogous problems to which existing solutions can be applied, or which suggest the existence of hidden correlations between factors (Davies, 1989, p.273). Certainly it is true that database technology and artificial intelligence have now both developed to the point that they can provide mutually useful support for the creation of knowledge-based systems.

The field of knowledge base management systems (KBMS) has arisen from the integration of established work in both database technology and artificial intelligence. To be termed a KBMS a system must contain a large database, an ability to make intelligent inferences from this database, and a satisfactory user interface. KBMS's have arisen from both the demands for sophisticated and easilly-accessible interfaces, and from the need to use artificial intelligence to provide smart strategies for navigating around and making inferences from the database.

One manifestation of this development is the way in which database systems and artificial intelligence are moving towards each other. In database research and development there is now a feeling that it is possible to go beyond the relational paradigm. This paradigm has been the dominant one since about 1970. It put forward the aim of representing information as the user wished to visualise it, i.e. as tables, rather than as the computer represented it (via pointers, arrays, and so on). However, the success of representation schemes from elsewhere such as semantic nets and semantic modelling from A.I. and the need for more complicated data structure representation such as hierarchies and hypertext have begun to give rise to demands for more advanced representation methods.

6.1.1. The ANSI/SPARC database standard

The last 20 years has also seen the successful development and evolution of the American National Standards Institute ANSI/SPARC model of database organisation. This envisages that databases be organised in three levels. At the top level there is the external layer of application programs, both in batch and interactive form. This includes the user sitting at his interactive terminal querying the database via some standard query language such as SQL. It is essential that many users be able to simultaneously access the database, so this requires some management policy on reading from and writing to the database in order to ensure consistency. At the middle level there is the conceptual view which is an abstract representation within the software using some conceptual data definition language and providing for mappings to the levels above and below. The bottom or internal level is the physical representation of data on disks, and is defined by an internal data definition language.

6.1.2. Expert systems

Expert systems made their presence felt in molecular science first in systems for selecting experimental procedures and analysing results (for example the Laboratory Information Management System LIMS), and for reaction searches (for example LHASA), and it is only comparatively recently that their potential has been realised for data retrieval and process control. In reaction pathway research (where the research thrust has been easy to maintain due to the vast resources of the drug companies) sophisticated knowledge bases have been created which contain information about pathways in the form of graphs. Topological reasoning can be performed on these graphs, where partial orderings are represented by the relation 'subgraph-of' (Wilcox and Levinson, 1985). Starting materials and reaction rules can be transformed into axioms, and automated theorem-provers can be used retrosynthetically to derive the target compound (Wang *et. al.*, 1985). These sophisticated knowledge bases represent a rapidly maturing AI technology applied to molecular science in which the classic knowledge representation problems have already been encountered (though not necessarilly solved) and for this reason hold much useful information for any researcher developing an expert system for protein structure prediction (Wipke and Dolata, 1985). Moreover workers in the field are now beginning to deal with the problem of linking the LHASA expert system to the ORAC reaction database. One of these problems is the interpretation of a chemist's sketch of the structural diagram, (Trindle, 1985). Besides the developments in database management systems (DBMS) artificial intelligence has also established new benchmarks of achievement together with a more tempered and realistic view of what AI can achieve within the foreseeable future. Knowledge based systems (KBS) have shown themselves able to successfully represent knowledge in many subtle ways, to infer using logic, to handle uncertainty, and to provide explanation to the user. What they are not in any sense is 'humanlike' or 'thinking machines'. KBS's represent only what has been put into them, and often this knowledge is quite shallow, sufficing for a wide range of standard situations but breaking down in an increasing number of non-standard ones, and requiring corrective maintenance in a continual attempt to attach new ad hoc algorithms and facts to subsume the widening field of relevance. It has been realised that what has been lacking has been any sophisticated way of representing the subtle and contingent nature of human knowledge. This has led to an increasing emphasis on 'user modelling', that is, representing not only the user's 'knowledge', but also a wide range of factors which cause this to change or to be interpreted differently. Within organisations such factors include socio-psychological and political considerations. Within scientific research, such factors include the criteria which cause different branches of disciplines to rise and fall as means of gaining funds and other political advantages, and the user preferences which influence how 'interesting' a fact is, and how 'relevant' one fact is to another fact. Thus increasingly it is being realised that while the expert may be indispensable for overall expert system design, the system should be flexible enough to allow the user to modify it over time, by means of inputting concerns such as preferences of relevance to himself.

Such considerations bring to the fore the differences between a DBS and a KBS. A database system (DBS) is primarilly concerned with implementation issues, concentrating on how the machine sees and processes the knowledge. This allows a clear distinction between abstract data structures and particular data instances. On the other hand, a knowledge base system (KBS) concentrates on how people see and process the information. This is often in a highly abstract form and so in such a KBS system there is no clear distinction between such general knowledge (perhaps represented as rules) and particular pieces of information (which may themselves be expressed as rules). The act of making distinctions between the abstract and particular in such systems has to involve some presuppositions as to the semantic meaning of the knowledge.

6.2. ELEMENTS OF DATABASE SYSTEMS (DBS)

6.2.1. The Relational Approach

Usually one begins specifying an information system where some restrictions have been accepted on the form in which information is represented. If such information is imagined to be in the form of a table, then the following restrictions would probably be acceptable to most of us:

(1) No two rows should be identical, so that we can specify a row by giving the values in that row.

(2) We can interchange any two rows without affecting the information content.

(3) We can interchange columns as long as we take column headings too.

(4) Only one value can exist in any box, where a box is defined as the intersection of any row with any column.

Such a system is said to be 'normalised'.

Another consideration in specifying an information system is the problems of repeated values within tables. Consider the following example.

SCIENTIST	Scientist_number	Group_number	Compound_number
	s_134	g_45	c_6543
	s_564	g_32	c_2754
	s_78	g_25	c_1998
	s_134	g_45	c_2754
	s_564	g_32	c_8776

We see that scientist s_134 appears twice, so that we can delete the second occurrence of laboratory g_45 without losing information (because scientist_number determines Group_number). This 'duplication' is not a problem : we can delete repeats without losing information. However, we see also that there are two occurrences of c_2754. This time however we cannot delete one of them because there is no clue within the table as to the information that has been deleted: i.e. we cannot delete entries without losing information. This occurrence of 'redundancy' is a problem which has to be got rid of by a method other than data deletion. The strategy is to split the information into subtables:

SCIENTIST/GROUP	Scientist_number	Group_number
	s_134	g_45
	s_564	g_32
	s_78	g_25
	s_134	g_45
	s_564	g_32

SCIENTIST/COMPOUND	Scientist_number	Compound_number
	s_134	c_6543
	s_564	c_2754
	s_78	c_1998
	s_134	c_2754
	s_564	c_8776

This process of detecting and removing redundancy is often called 'normalisation'.

Consider now a completely different problem described by the following statement

> A scientist is uniquely identified by a number. There are several scientists in each laboratory. A project may involve several scientists from several laboratories. Each compound has a unique number. Each compound may be mentioned in several patents. Each scientist uses several compounds.

This information could be represented inside a database system as follows

```
------------------------------------------------------------
        SCIENTIST ----> LABORATORY
        SCIENTIST ----> COMPOUND
        PROJECT   ----> SCIENTIST
        PATENT    ----> COMPOUND
------------------------------------------------------------
```

where arrows indicate the relation 'determines'. For example, if we are given the identity of a scientist we can know which laboratory he works in, and if we are given a project number then we can say which scientists are involved in it. In order to avoid redundancy each of these determancies needs to be expressed as a separate subtable, except that we can have a subtable that includes SCIENTIST,LABORATORY, and COMPOUND. Thus this information system generates three subtables, SCIENTIST-LAB-COMPOUND, PROJECT-SCIENTIST, and PATENT-COMPOUND.

The above specification method is called the 'bottom-up' method, by looking at tables and subtables. The alternative method is the 'top-down' method which involves finding 'entities', their 'attributes', and the 'relationships' which connect them. For example, for the previous problem, the list of entities and their attributes and relationships would be something like this:

ENTITY	ATTRIBUTES
Compound	compound-number, property-list
Scientist	scientist-number, scientist-name
Laboratory	lab-number, telephone-number
Patent	patent-number

RELATIONSHIPS		
Scientist	uses	Compound
Scientist	located-in	Laboratory
Patent	mentions	Compound
Project	employs	Scientist

Information about relationships is merely sufficient to answer yes or no to such questions as 'is scientist X located in laboratory Y?'. Entities have one attribute which is the 'identifier' for the entity, such as scientist-number for entity Scientist. Attributes cannot appear in more than one entity, though they can appear in combination with each other in what are called 'compound entities' (see below). This description glosses over many difficulties and alternative solutions (for example, the above formulation means that it is not possible to answer the question 'which laboratory did scientist X use compound Y in?'. Such questions can be answered using three-way relationships, or giving relationships their own attributes ('uses' could have an attribute 'located in' and we could dispense with the relationship of that name), or we could create compound entities such as Materials whose identifier is a combination of scientist-number with compound-number with laboratory-number.

Despite criticisms of the relational paradigm there is a massive current commitment and much resistance to change. For example, the STandard for Exchange of Model data, (STEP) currently under development for the ISO committee has decided not to make any change from the relational approach. One reason for the delay in change from relational databases is the massive effort expended on extending them to cope, for example, with complex objects and operations, and with logic programming constructs as in the Algres system (Ceri et. al., 1990). It seems most likely that there will be a transitional period during which vendors provide non-portable relational products until consumer frustration with non-portability reaches a critical peak (Stonebraker, 1989) when a completely new type of database becomes available.

6.2.2. Object-oriented databases

In the last few years new ideas from software engineering on how to construct programs to ensure reliable functioning have begun to influence database design and to undermine confidence in existing methods. One of these ideas from software engineering has been that of object oriented programming. With an object-oriented database data can be stored as an abstract object. The database does not force the user to convert the information into facts and figures that will fit neatly into a table. An object may subsume many other objects. The database will retrieve all of these objects as a whole. Abstract relationships between these objects can be stored, and similar objects can be clustered together. This feature is useful for a biologist wanting to search a database for a particular class of sequences. The data representations available are potentially very various indeed. So data can be represented as strings of characters, as digitised images (where pixel coordinates are given values of 0 or 1), or as tables. This flexibility of representation and ability to group according to salient features makes object-oriented databases potentially more attractive than relational ones.

These features have resulted from programming languages which have object-oriented features. In an object oriented program there is a hierarchy of modules. Each module explicitly imports and exports information via 'messages' from and to other modules. Each module has knowledge only of that information in other modules in the program which has been explicitly declared to be available to it. All the inner workings of other modules are hidden to it. Also the objects which are created have certain characteristics or 'types' (for example, one object-type may be 'display-self'). These characteristics can be declared in a very general and abstract way such that more specific meanings are given to them when they are actually implemented within a specific module. And objects can belong to classes which share characteristics, so that modules can inherit quite complex attributes from the class to which they belong. So a module TEXT might be able to inherit 'display-self' type from module GEOMETRIC_OBJECT, if both modules are in the same class. Such a view of program development and construction has been shown to give rise to much clearer and safer programs. The development of new object oriented languages such as ADA and SMALLTALK has been important in view of the fact that something like threequarters of all delivered software develops bugs which often lead to system failure. But also these ideas have had a deep effect on forcing a rethink on how databases are designed and built, particularly with regard to the need for databases to retain integrity during multiple reads and writes, with regard to the greater need for security, and also with regard to the huge dependence which modern industry places upon databases for its smooth functioning. The interest in object-oriented databases is just one example of this new thinking about how programs are written.

The use of object-oriented methods in database applications has partly been motivated by a realisation that the benefits of redesigning the query language and its interface with the programming language defining and manipulating the database would be small compared with the benefits of redesigning the programming language. This has led to attempts at designing languages which have the functionality of a fully-fledged database management system as well as that of a programming language, and which are also object-oriented (Dittrich, 1988, Schmidt et. al., 1988).

6.2.3. Semantic Databases

The other major programming development in the last ten years of importance to databases is the growth of semantic modelling (Hull and King, 1987). Traditional relational database systems have used the record as the principal data representation medium. While in the older non-relational systems the order of items in tables and other structures was significant, in relational systems the order of tuples (ordered sets) within tables is not relevant to the user: he is only interacting with the system at the level of description of 'tables'. Relational databases thus have used the record (a row in a table) and the order of fields within all the records has been hidden from the user. Traditional relational systems have used only simple data types for fields: numbers and strings. In contrast, the development of semantic databases has involved the construction of a rich set of abstract types and methods for representing them. The most abstract type is the 'entity type', such as SCIENTIST or PATENT. Occurrences of entity types are 'printable types' such as SCIENTIST_NAME or PATENT_NUMBER, which are printable in the sense that Dr. Jones or Patent number 0106452 are instances of them.

Entity types may have subtypes: for example, PATENT may have subtypes BIOCHEMISTRY-PATENT , BIOTECHNIOLOGY-PATENT, and ORGANIC CHEMISTRY-PATENT. Such subtype-supertype relationships are usually called 'ISA' relationships. For example, each biochemistry patent 'is a' patent. Subtypes can be 'user-specified' or they can be 'derived', the latter referring to aspects of the object which allow the system to deduce that the object fits a particular subtype. Thus scientist Dr. Jones may be of user-specified subtype BIOCHEMIST-SCIENTIST while Compound X may have attributes which allow the system to give it the subtype DANGEROUS-COMPOUND. Obviously the more automated the system can be the better, and so one always tries to maximise the number of 'derived' subtypes. Another important characteristic of semantic modelling is the way in which composite objects can be created from atomic objects. For example, type COMPOUND can be constructed from the types ELEMENT, TEMPERATURE, PRESSURE, and so on. Another important tool within semantic modelling is the use of sets created by association. For example, one such set might be all patents written by scientist Dr. Jones, or all dangerous compounds used by Dr. Jones for experiments which resulted in patents. Attributes of objects are represented by functions (arrows, pointers, or some other representation method). So for example, the subtype NEW-COMPOUND-PATENT might have the attribute HAS-NUMBER mapping to PATENT-NUMBER, the attribute DISCOVERY-BY mapping to SCIENTIST-NAME, and so on.

6.2.4. Object-oriented and semantic databases compared

Object oriented databases and semantic databases have retained separate identities and have tended to serve distinct purposes. Semantic modelling serves the structural aspects of systems, such as data types. Thus in semantic databases classes of types and inheritance of type attributes by subtypes are the principal abstraction medium. On the other hand in object oriented databases it is not the structural features of objects, but their dynamic and behavioural aspects which are highlighted. Thus the object type SEQUENCE may have the method 'compare for similarity', which can be inherited by the object type STRUCTURE, a seemingly unlike type. This is markedly different from semantic databases in which the inheritance of attributes can only take place between types where one is a subset of the other.

6.3. ELEMENTS OF KNOWLEDGE BASE SYSTEMS (KBS)

Knowledge base systems differ from database systems in fundamental ways. The facts and their interpretation are usually much more systematically organised according to a theory and in such cases rules and predicate logic can be used. For example

```
----------------------------------------------------
FACT BASE
----------------------------------------------------
compound1(compound-number)
compound2(compound-number)
compound3(compound-number)
......................
----------------------------------------------------
RULE BASE
----------------------------------------------------
compound1 implies danger
compound2 implies need-compound3
compound4 implies use-twice
compound5 implies danger
danger and use-twice and not compound2
          implies extra-scientist-needed
experiment1 implies compound1 and compound2
......................
----------------------------------------------------
```

Such a rule-based system might be used to discover the various compounds and conditions needed to carry out a series of experiments.

Semantic nets are also useful in knowledge base systems. They are made up of nodes and arcs (arrows between nodes). Nodes represent ideas or things (real world objects). Arcs represent relationships between things. Relationships are usually ones of generalisation, specialisation, aggregation, and classification. Examples include 'is a' in 'man is a animal' and 'a kind of' in 'dinosaur a kind of toy' and 'a part of' in 'leg a part of man'. Classification typically involves time, subject, object,place, motive, and so on. Semantic nets allow more complex and subtle systems to be built but are also more difficult to construct.

Another technique developed within artificial intelligence and of relevance to the development of knowledge-based management systems is that of frames. Frames are containers with slots to hold different pieces of information about an instance of a subject. So an example of a frame might be 'dangerous experiment' and one instance might be 'dangerous experimentl'. It might be that no information exists about experimentl apart from its dangers. In this case the information contained in the slots of the general frame 'dangerous experiment' would be used by default.

6.4. ELEMENTS OF A KNOWLEDGE-BASED MANAGEMENT SYSTEM (KBMS)

Knowledge based management systems involve combining the advances made in both database technology of DBS and the artificial intelligence technology of KBS. Typically such systems aim (although they do not always achieve their aims) to provide the following advantages over more traditional approaches:

6.4.1. Dynamically updated knowledge bases

Artificial intelligence programs acquire new knowledge and add this to their knowledge base. Such new knowledge can come from making deductions from information supplied, but more importantly, it can come from asking the user for a crucial and previously-withheld fact. Such new information must be consistent with old information. If it is not the conflict must be resolved into some new concept or the user must be asked for help in resolving the conflict, perhaps by asking him to expand the system's underlying concept base.

6.4.2. Natural language interfaces

Natural language processing is a relatively new field. One of its goals is to enable the user to communicate with the computer in a language that he is familiar with, ie in natural language. It involves inputting text strings. Once this goal has been achieved completely then it will be possible for research to concentrate on speech processing. Directly addressing the computer using speech which does not necessitate learning any new language can be seen as the ultimate goal. One would speak directly to the computer and the computer would provide an answer, another question, or would say it could not provide an answer. Communication with the computer has its own problems in addition to those posed by text processing. Eventually such speech input systems will become commercially available, but it is difficulat to say when this will be.

One problem of such systems is that they are only feasible when dealing with a restricted grammar and vocabulary, and are thus only useful for restricted applications: for doctors querying patients' symptoms, or for mechanics querying car problems, for example. This problem is described as 'domain dependence', that is, the language set is applicable only to one particular problem situation. McDonald and Yazdani (1990, p.176) comment:

> "Natural language is loose, vague and ambiguous, consequently natural language systems are large and unwieldy, with overheads on efficiency making a full system extremely slow. Language is dynamic, in a constant state of flux, therefore natural language processing is likely to remain a problem for some time. What is required is a subset of natural language that facilitates communication in all domains, but a useful subset is a long way off; commercial subsets tend to be domain dependent."

An example of a system which uses a natural language interface is UM-tool, which is a knowledge-based system which creates, maintains, and modifies user-specific models of the user's needs within an expert system context. Its knowledge domain is search strategies in bibliographic databases (see also Smith and Chignell, 1984, Walker and Janes, 1984, Clark, 1984). It can be interfaced with another piece of software for searching bibliographic databases. Here is a representative session between a user (U) and the system (S):

"S: Please, state your information needs.
U: I am interested in papers dealing with computer assisted NMR devices.
S: OK. I understand the following concept: Cl: Computer-assisted NMR Devices. Is that correct?.
U: Yes.
S: Please, select any of the following term combinations if you think they express the same meaning as concept Cl:
Computer and NMR Medical Devices
Computer Devices
Computer-assisted and NMR Medical Devices
Computer-aided and NMR Medical Devices
Computer-based and NMR Medical Devices
U: Computer and NMR Medical Devices is the best combination, but also the others are good ones, except for cComputer Devices.
S: Now please choose the discipline(s) related with the topic(s) of your query and rank them (1-5). The scale is
5 very relevant
4 relevant
3 not known
2 hardly relevant
1 not relevant

The disciplines are:
Administration
Aeronautics
Agriculture
....

U: Computer Science (5) for Computer and Medicine (4) for NMR Medical Devices.
S: OK. Please, tell me why you need this information (to write a report, to prepare a thesis....)
U: I am going to write a course for operators of specialised medical machines.
S: I understand the following:
Your information need derives from the fact that you have to teach a course. Its topic is about operators of specialised medical machines.
U: No. The topic is specialised medical machines.
S: I understand the following:
The course topic is specialised medical mchines.
Is this correct?
U: Yes.
S: Considering your previous search sessions and the disciplines involved in the current search, I would suggest the following databases:
INSPEC, NTIS, PASCAL"

(Brajnik et. al., 1990).

In this example, stereotypes are used, and classification is dynamic. Information input is question and answer, and this is used by the system to build up a complex model of the user: his level of subject knowledge, his search sophistication, his purpose in searching, and so on. The example reveals the difficulties of state-of-the-art applications: their time-consuming nature, the reliance on a slow and difficult keyboard rather than speech, the rigidness and inability to take shortcuts. GenInfo, the experimental retrieval system developed by the US National Library of Medicine to assist biologists in accessing a large collection of different molecular biology factual databases, suffers from similar problems. It seems likely that future applications will show many improvements.

The user should be able to type in strings which are parsed for semantic meaning by the system. At the present state of technology this is possible but in a limited way. For example in the library system discussed above, the system grabs word strings containing keywords and prompts to make sure it has deduced correctly what this string means. The result is a semi-natural language dialogue which is better than a structured query language, but which still has the disadvantage of being slow due to the high frequency of requests for the user to confirm the system's deductions. Until speech synthesis becomes feasible such an approach is inevitably going to be slow. Also this approach requires some way in which the user can communicate his impatience or his requirement that the system proceed faster, very much as humans do. Such aspects of user interface design are becoming more important now. An alternative to natural language processing is graphical query systems with rapid mouse clicking. They too have their weaknesses. Again there is the problem that the number of menus may become burdensome, especially for an advanced user, but also even for a novice, who after all is probably less likely to foresee the advantages of proper use of the system. More generally, a software layer is required for applications interfacing to provide specific functionalities for different user needs. These include: image retrieval and presentation, text processing and editing, form management and utilization, document entry and composition, complex objects entry and retrieval, computer-aided design and simulation for laboratory and industrial applications, and expert system interfaces.

6.4.3. User modelling

One of the current problems with current expert systems is their homogeneity compared with the diversity of user needs even within the same knowledge domain. A necessary means of alleviating this problem is the ability of any system to ascertain the user's preferences, together with rules for determining how to proceed based upon these preferences (White, 1990). Such rules are most likely going to use methods which have been around for some years within statistical decision theory. For example, multi-attribute decision analysis involves scoring alternatives on relevant dimensions and choosing the alternative which has the largest score, after allowing for weighting and other modifications.

6.4.4. Self-modifying consistency checks

In a large system involving values and interpretations together with fuzzy and qualitative reasoning, the scope for misinterpretation and ambiguity are enormous. The problem of the system designer is how to build in sufficient checks with the user without overburdening him with queries. This must inevitably involve constructing theories of consistency and correctness which can become self-modifying in some way. If this is not done then the user is continually prompted for information and guidance whenever ambiguity arises. This requires advances in the representation of such aspects, and also in the modelling of human qualitative reasoning, which is not well understood. This problem is just as important in scientific areas as elsewhere. Although scientific facts are unambiguous, the process by which clever scientists guess that gaps exist or that facts can be associated in new ways means that the arrangement of facts and their associated inferences is neither unambiguous nor fixed. The problem is that facts must be represented in a contingent, temporary status, allowing sufficient flexibility for changes to be made in the inferences that can be drawn from them and at the same time the system must be able to tolerate the ambiguity thus generated.

6.4.5. Knowledge representation

The area of knowledge representation is and has always remained one of the most difficult problems in artificial intelligence. It has been shown that many problems in AI depend upon the adequate representation of knowledge for their successful solution. For example, the increasing importance in molecular science of three-dimensional information for the understanding of protein functioning means that such information must be adequately represented and easilly accessible. Comparison of the three-dimensional attributes of parts of two proteins, for example, might be the kind of representation needed. The fifth-generation project, begun in 1981 by Japan and quickly followed by Europe and the United States, is aimed at parallel hardware, applied A.I., and new software systems. Databases are an integral part of the project, and one of the database component's goals is the development of methods of representation of complex objects comprising text, graphical images, and digital images. This requires a software layer which transforms complex objects into their different forms: pictures described by text and so on.

6.4.6. Knowledge manipulation and retrieval

How knowledge is manipulated becomes important when considering the vast size of modern scientific databases and the potentially long times for searches to be carried out. Human memory is accessed on the basis of some echo of relevance or bright colouring which is able to attract the remembering person's attention among the vast cloud of memories competing for retrieval. Although research has been carried out on content-addressable memory for computers this is not a feasible hardware option in the foreseeable future. Eventually it is expected that a multi-media associative memory will become available to provide storage for complex objects (text and pictures) using associative memory and parallel processing. However, this is still a research project at present. Thus one is forced back on the traditional means by which machines refer to information: each piece of information is assigned to an 'address' in memory. It can only be retrieved by giving the correct address. Thus information content has no connection with where it is stored. It seems likely that some human attributes can be mimicked. For example facts could be continually rearranged within the machine's memory according to certain geographical rules - each type of fact would have its own territory. Also it may be possible to mimick content-addressability in software using neural nets (which allow vast numbers of interconnections to be created between facts - such as all facts related to 'danger' being connected to a 'danger' node).

6.4.7. Reasoning

Any KBMS must include a reasoning capability. Usually such a capability means that the system can use propositional and predicate calculus. An example of propositional calculus is the rule of modus ponens ponens which says that, if P and Q are truth-statements, given that P implies Q, and P are true then we can deduce the truth of Q. An example of predicate calculus is the rule of universal quantifier elimination which says that from the statement ' for all members of a particular set, P is true' we can deduce the statement 'for any member of that particular set, P is true'. From these apparently simple and transparent rules we can determine the truth or falsity of very complicated logical expressions. It will probably come as no surprise to the reader that such computations are rapidly solvable by a machine. Even vagueness and lack of information can be modelled by means of probabilities and by using 'fuzzy reasoning'. Probabilities can be used to represent levels of certainty (near 1.0) or uncertainty (near 0.0). However it has been found that in practical systems often the principles of statistical theory are contravened. For example, probability theory assumes that the events under consideration are unrelated, yet in reality such a principle is always infringed. Another approach is to use 'fuzzy reasoning' where instead of there being a rule for what happens if a condition is met and for what happens if it is not met, we also have a rule for what happens if we are not sure that the condition is met. There are some detractors from these approaches to logic. One of their main criticisms has been that if such rules were written in a list then they would be unable to accept any revision or rethinking such as humans make in reality. This is because such an approach to logic is 'monotonic', which in this context means that if we have at a particular point in a program decided that P is true, then we cannot re-evaluate that proposition later and decide that P is false. This has given rise to 'non-monotonic' logics which are always open to revision : essentially what occurs is that if two propositions exist in the program 'P' and 'not P', then the program takes the later of the two contradictory propositions as the correct one.

6.4.8. Using reasoning to update the knowledge base

A knowledge base which is created at the start and is never updated is not useful for dynamic or complex problems where new sources of information or new goals are being discovered. Such situations require an 'open eyes' approach in which new information is continually collected and existing knowledge revised and reorganised (Bachman, 1989). This requires some means by which the knowledge base updates can be automated, since the continuous use of experts is expensive and error-prone. Some attempts have been made recently to develop methodologies that permit chemical expert systems to work directly with raw factual chemical data in order to eliminate manual analysis of the information and also to eliminate manual updates. Such attempts have involved some reasoning capability for the derivation of the rules which can be said to summarise the information contained in supplied 'training examples' embedded in the raw data (Wipke 1989).

6.4.9. Explanation

Another facility which would be required within a KBMS is that of explanation. Questions of the type 'what does this fact lead to?' can be answered by the system 'forward chaining' from a condition to its consequences. Questions of the form 'why do you say that?' are answered by 'backward chaining' backwards from consequences to all conditions which could have acted as a trigger. The user must be able to inspect the system's reasoning and satisfy himself of the correctness or otherwise of such reasoning. Systems have started to be built which adapt the complexity of their explanations to the sophistication level of the user. Further advances are necessary in the direction of making alternative types of reasoning (see above) available within constrained regions, perhaps as an experiment to show the user what the alternative inferences are to a set of facts.

6.4.10. Security

It is just as important within a KBMS for there to be adequate methods of ensuring security against uncontrolled access. Indeed, the intelligence capabilities of the KBMS would be useful weapons in alerting authorities to such dangers, and in monitoring non-permitted use.

6.4.11. Integrity

Because it is possible that unpredictable writes will be made to the knowledge base (in reaction to new information from the user or in reaction to new deductions made by the KBMS) it is difficult to ensure data integrity since an earlier answer and a later answer may be inconsistent. Such inconsistencies need to be monitored automatically and the user alerted to any adverse implications.

6.4.12. Protection

Database systems always have the potential to breakdown due to system software failure or due to hardware failure. In such cases it is of the utmost importance that the database be protected so that no data is lost or corrupted by overwriting.

6.4.13. Pictorial databases

The requirement to keep pictorial data enlarges the requirements field by specifying various intelligent strategies for querying such information. As in most present-day applications, the prime requirement is for interactive processing, together with methods of simplifying images when the user is interacting with the interface (e.g. the user is given two ikons as alternatives to choose from). An indispensible part of a pictorial database is a 'summary information' database, which categorises and indexes the images, together with text descriptions (Crehange et. al., 1984). Such summary information is very poor compared with the information contained in the images, and therefore contains incomplete and imprecise information. The summary information consists of dimensions. These include a description of what is visible in each image, possibly ideas resulting from them, details about events when the image was created (e.g. experiment and its purpose and result), and so on. A database management system is not by itself adequate to deal with such a problem. The role of any expert system in such an area is therefore in providing 'fuzzy' recognition methods. The task can be broken into subtasks: requesting information from the 'summary information' database, aided by image-simplification methods, possibly altering these requests along certain dimensions (e.g. altering beta to alpha, altering angles), sampling images, creating 'archetype' images (both by the user and by the system and by both together interactively) which contain common elements from a group of images, and so on.

6.4.14. Moving images

In the case of moving images involved for example in reactions, the pictorial database will contain sequences of images, and it is the recogintion of sequence attributes which the expert system aims at detecting. One problem involved in such cases is the need to say when one process has finished and another one has begun (Walter et. al., 1987). This expert judgement needs to refer to the 'summary information' database using rules for decomposing sequences into semantically meaningful processes.

There are various ways in which these aims can be achieved within a KBMS. One approach is to attempt to keep the AI and database aspects within one single system, using one single programming language. An example is the programming language PROLOG. However, due to its origins in logic programming this approach is always going to favour the inference engine part rather than the database management part of the problem. Such an approach therefore necessitates specific tailoring of the system, including tacking on database management capabilities. Another approach is to put together two distinct systems, one an expert system, the other a database manager. Such a double act can be a loose one, where the expert system occasionally takes snapshots of the database, or it can be a tight one, where the data in the database is available at any time to the expert system. The problem with the latter approach is that the interface between the two systems will inevitably cause problems over time. A third approach is the most revolutionary. This is to rethink the nature of database and of knowledge-based systems, even to the point of creating new purpose-built languages more suitable for doing the job.

The area of KBMS is still undergoing active research, and systems developed within the KBMS area (Kerschberg 1986, 1988, Schmidt and Thanos, 1988, Ullman 1988, Smith 1986) are as yet prototypes rather than commercial systems.

6.5. REFERENCES

Bachman C.W., 1989, "A personal chronicle: creating better information systems, with some guiding principles", IEEE Transactions on Knowledge and Data Engineering, 1, 1, 17-32.

Brajnik G., Guida G., and Tasso C., 1990, "User modelling in expert man-machine interfaces: a case study in intelligent information retrieval", IEEE Transactions on Systems, Man, and Cybernetics, 20, 1, 166-185.

Ceri S., Crespi-Reghizzi S., Zicari R., Lamperti G., and Lavazza L.A., 1990, "Algres: an advanced database system for complex applications," IEEE Software, 68-78, July.

Clark W.B., 1984, Supplantation of cognitive processes as a user aid in online systems, 106-109 in Challenges to an information society. Proceedings of the 47th ASIS annual meeting,

Crehange M., Haddou A.Ait, Boukakiou M., David J.M., Foucaut O., and Maroldt J., 1984, EXPRIM: an expert system to aid in progressive retrieval from a pictorial and descriptive database, 43-61 in Gardarin

G. and Gelenbe E., (Ed.), 1984, New applications of databases, Academic Press, London.

Davies R., 1989, "The creation of new knowledge by information retrieval and classification", Journal of Documentation, 45, 4, 273-301.

Dittrich K. R., (Ed.), 1988, Advances in object-oriented database systems. 2nd international workshop on object-oriented database systems, Bad Munster am Stein-Ebernberg, FRG, September 1988, Proceedings, Lecture Notes in Computer Science Vol. 334, Springer Verlag.

Kerschberg I., (Ed.), 1986, Expert database systems, Proc. 1st Int. Conf., Benjamin/Cummings, Redwood City, CA.

Hull R., and King R., 1987, "Semantic database modelling: survey, applications, and research issues", ACM Computing Surveys, Vol. 19, No. 3, September,201-260.

Kerschberg I., (Ed.), 1988, Expert database systems, Proc. 2nd Int. Conf., Benjamin/Cummings, Redwood City, CA.

McDonald C. and Yazdani M., 1990, Prolog programming: a tutorial introduction, Blackwell,London.

Morris A. and O'Neill M., 1988, "Information professionals - roles in the design and development of expert systems?" Information Processing and Management, 24, 2, 175-181.

Schmidt J.W., and Thanos C., 1988, Fundamentals of knowledge base management systems, Springer-Verlag, Berlin.

Ullman J.D., 1988, Principles of database and knowledge base systems, Vol. 1, Computer Science Press, Rockville, Maryland.

Schmidt J.W., Ceri S., and Missikoff M., (Eds.), 1988, Advances in database technology EDBT '88. International Conference on extending database technology. Venice, Italy, March 1988. Lecture Notes in Computer Science Vol. 303, Springer Verlag, Berlin.

Smith P.J., and Chignell M., 1984, Development of an expert system to aid searches of the Chemical Abstracts, 99-102 in Challenges to an information society. Proceedings of the 47th ASIS annual meeting,

Smith J.M., 1986, Expert database systems: a database perspective, Benjamin/Cummings, Redwood City, CA,USA.

Stonebraker M., 1989, "Future trends in database systems", IEEE Transactions on Knowledge and Data Engineering, 1, 1, 33-44.

Trindle C., "An intelligent sketchpad as a medium for chemical information", Chemical Information Bulletin, 37, 2, 31.

Walker G., and Janes J.W., 1984, Expert systems as search intermediaries, 103-105 in Challenges to an information society. Proceedings of the 47th ASIS annual meeting,

Walter I.M., Lockemann P.C., and Nagel H-H., 1987, Database support for knowledge-based image evaluation, 3-20 in Stocker M. and Kent W., (Eds.) 1987, Proceedings of 13th international conference on very large databases,Brighton Morgan Kaufman, Los Alamos, USA.

Wang T., Burnstein I., Ehrlich S., Evens M., Gough A., and Johnson P., 1985, "Using a theorem prover in the design of organic synthesis", Chemical Information Bulletin, 37, 2, 35.

White C.C. III., 1990, "A survey of the integration of decision analysis and expert systems for decision support", IEEE Transactions on Systems, Man, and Cybernetics, 20, 2, 358-364.

Wilcox C.S., and Levinson R.A., 1985, "A self-organising knowledge base for recall, design, and discovery in organic chemistry", Chemical Information Bulletin, 37, 2, 35.

Wipke W.T., 1989, "Analogical reasoning from chemical knowledge bases", Chemical Information Bulletin, April, 26.

Chapter 7:

The Main Sequence Databanks in Molecular Science

7.1. DEFINITIONS

The term databank in the broad sense means a machine-readable collection of factual information which is often referred to as a non-biliographic database or numeric database. In other words a databank is a "collection of numeric, textual, or factual information which minimises the need to access original sources" (Ash *et al.*, 1985) such as journal articles, handbooks, textbooks, theses, and reports.

Data such as the numeric representations of the magnitudes of various quantities can also include specific but non-numerical scientific facts such as structures of molecules or biological properties. The word "databank" is not used consistently in many cases probably because most of the databanks in molecular biology for example have bibliographic references as well as numeric information included in them.

Sequence databanks include the "linear script of nucleic acid and protein sequences which in principle tells all there is to know about the genetic makeup of an organism - its genetic program - and also contains vestiges of the evolutionary history of the species", (von Heijne, 1987). It is of great importance that each database's store of molecular information should be as accessible as possible. Von Heijne's description not only gives a definition but indicates the main objectives of a databank and its potential importance which is a function of use. The use to which databanks are put is dealt with in the following chapters.

7.2. SHORT HISTORY OF SEQUENCE DATABANKS

Although the complete amino acid sequence of the beta-chain of bovine insulin (a short polypeptide of 30 residues) was determined as early as 1951 the method was slow and there were only 27 sequenced proteins (mainly polypeptides such as haemoglobins and cytochromes) in the first "Atlas of Protein Sequence and Structure" compiled by Dayhoff and her colleagues (Dayhoff, 1965).

More than ten years later the complete nucleotide sequence of transfer RNA from yeast was published (Holley et al., 1965) and was included in the next edition of the "Atlas" in 1972 which contained over 400 protein sequence entries with a few nucleic acid (RNA) sequences (George et al., 1988).

With improved protein purification methods (Schroeder, 1968) and a "breakthrough" in sequencing methods by Maxam and Gilbert (1977, 1980) and Sanger et al, (1977) the number of known amino acid sequences increased rapidly. New methods in genetic manipulation made it possible to determine amino acid sequences from nucleic acid sequence which not only increased the amount of data exponentially but also improved their quality and changed their scope (for example, not only soluble proteins can be dealt with, as had previously been the case).

Collections like the one at the U.S. National Biomedical Research Foundation (NBRF) and similar ones at other research laboratories had grown so fast that it became necessary to store, retrieve, and analyse data using online facilities. Even this type of data storage has limitations and so there were improvements, especially now that the determination of the human genome sequence with 3,000 million nucleotides is becoming a probability despite difficulties of research funding.

It should be stressed that at present only 0.2-0.3 percent of the human genome is known. An individual scientist can only determine and treat 10 kilobases (a kilobase is 1000 base pairs) per year. However, with the adoption of new automated techniques and improvements in work organisation this rate will change to about 100-200 kilobases per year per scientist. In 1989 the volume of data already incorporated in the set of all databanks was about 20,000 kilobases. At present these databanks can acquire up to 10,000 kilobases per year (see Figure 7.1).

The nucleotide set of a typical micro-organism such as E. coli would be 4,700 kilobases, equivalent to a book of 1500 pages if the set were to be printed out. In contrast the human genome set would occupy a book of one million pages.

In terms of numbers of protein sequences, in 1988 there were around 12,000 protein sequences held on databases, and new ones were being entered at the rate of 100 a month (Lesk, 1988, p.31). Between 1982 and 1989 the EMBL Nucleotide Sequence Database grew 40 times (Kahn and Cameron, 1990). It is expected that over the next decade biomolecular databases will grow between seven- and sixty-fold (Colewell, 1989).

Although the main databanks such as NBRF, EMBL, GenBank etc, are described elsewhere, (Bishop et al., 1987; Lesk, 1988), there are constant changes taking place due to the growth of data and the need for better utilization of information generally through easy access and increased speed in data processing and retrieval. It is these aspects which will be discussed below. The relevant articles here are written by scientists who use the databanks and who therefore are in a good position to write informatively about them. These articles are scattered

Genomes	Kilobases
theoretical minimum	1,000
Escherichia coli	4,700
Saccharomyces cerevisiae	15,000
Aspergillus nidulans	26,000
Arabidopsis thaliana	70,000
Human	3,000,000
Sequencing (man/years)	**Kilobases**
micro	10
macro	100-200
mega	6,000
Collecting	**Kilobases**
database total content 1988	20,000
database input/year	5-10,000

Figure 7.1. Comparison of data volumes in nucleotide sequences (Keil, 1990, p.51).
Reprinted with permission from Keil B., "Cooperation between databases and the scientific community", in Doolittle R.F., (Ed) "Molecular evolution: computer analysis of protein and nucleic acid sequences", Copyright (c) 1990, Methods in Enzymology, Vol. 183, Academic Press.

about in subject-oriented journals such as Nature, Nucleic Acids Research, Journal of Molecular Graphics, Trends in Biochemical Sciences, Methods in Enzymology, Proceedings of the National Academy of Sciences of the USA and so on. They are not mentioned very often in library or information science journals.

7.3. WHAT DATABASES ARE AVAILABLE ?

Figure 7.2 provides brief information on the content of online databanks for biological macromolecules. Some of them give information (literature, citation, sequence, annotation, genetic map) derived from different species (e.g. GenBank) while others collect specific data on one species only (e.g. OMIM).

Database	Type	Database Name and Data
AANSPII	3,4	(Amino Acid Nucleotide Sequences of Proteins of Immunological Interest) immunoglobins
AGRICOLA	11	(AGRICulture and Life sciences) bibliographic records for literature citations from 2500 journals
AIMB	11	(Artificial Intelligence and Molecular Biology) list of researchers
AMINODB	9	(Amino Acid Data Base) molecular properties of amino acids
BCAD	1	(BioCommerse Abstracts and Directory) worldwide news index
BIOSISCONN	11	(BIOSIS CONNection) data and literature citations in life sciences
BIOSISP	11	(BIOSIS Previews) machine-readable version of Biological Abstracts (BA) and BA/RRM (reviews, reports and meetings)
BKS	11	(Biotech Knowledge Sources) review of biotechnology publications
BMCD	6,7	(NIST/CARB Biological Macromolecule Crystalisation Database) crystalisation conditions for crystal forms of biological macromolecules
BMRB	6	protein NMR data
BRD	2	(Berlin RNA Databank) tRNA sequences
BRENDA	7	enzymes,kinetics
CAS	11	(Chemical Abstracts Service ONLINE) chemical literature abstracts
CASORF	various	(CAS Online Registry File) chemical structure and dictionary with more than 10 million substance records including structure diagram, molecular formula, 10 most recent citations, and 3D ccordinates (if applicable)
CATGENE	11	(domestic CAT GENE frequencies) gene frequencies and statistical analyses
CCD	10	(Cambridge Crystallographic Database) small molecule atomic coordinates
CGC	11	(Caenorhabditis Genetic Centre) C. elegans strains with genetic map and bibliographic data
CSD	5	(Carbohydrate Sequence Database) carbohydrate sequences
CSRS	2	(Compilation of Small RNA Sequences) small RNA sequences
CURRCONTS	11	(CURRent CONTentS) contents pages of 1200 life science journals
CUTG	3	(Codon Usage Tabulation from GenBank) codon usages in all available genes from GenBank
DBEMP	7	enzymes,kinetics,pathways

DBIP	11	information on peptides and proteins including literature citations, synthetic peptides and protein sequences
DBIR	11	(Directory of Biotechnology Information Resources) contact names and addresses in biotechnology
DCT	3	(Drosophila Codon Tables) codon usage by transposable element DRFJ nucleotide sequences: (host gene) nucleotide sequences
DDBJ	3	(DNA Databank of Japan) nucleotide sequences
DIALOGMC	11	(DIALOG Medical Connection) literature citations and abstracts
DRHPL	11	(Database for the Repository of Human Probes and Libraries) chromosomal assignment, literature references, cross-references for chromosome-specific libraries and human genomic and cDNA clones in Repository
DROSO	11	(Genetic Variations of Drosophila Melanogaster) machine-readable book "Genetic Variations of Drosophila Melanogaster"
ECOLI	2,3,4	(ECOLI k12 genome and protein database)
EMBL	3	(European Molecular Biology Laboratory) Nucleotide sequences
EMBOPRO	4	automatically generated amino acid data from EMBL database
ENZYME	11	(ENZYME database) catalytic activity,cofactors, and cross references to SWISS-PROT
EPD	3	eukaryotic POL II promoter nucleotide sequences
GBSOFT	11	(The GenBank Software Clearing House)
GC	11	(Gene Communications) guide of human genome clone availability, lists published reports of cDNA, genomic and synthetic clones comprising gene and pseudogene sequences, uncharacterised DNA segments, and repetitive DNA elements
GDB	various	(Genome Data Base) genetic mapping and disease data to support mapping and sequencing of human genome
GDN	11	(Gene Diagnosis Newsletter) guide to publications on human inherited diseases using recombinant DNA methods
GENATLAS	11	human gene map data including markers localised on chromosomes, gene maps of animal species
GENBANK	3	(GenBank Genetic Sequence Databank) nucleotide sequences, catalogued and annotated by site of biological significance
GENESEQ	3	(GENESEQ) nucleic acid and protein sequences taken from published patent information
GENINFO	various	(GENINFO) compendium of data in other databases with date saved
GRIN	11	(Germplasm Resources Information Network) data on 8000 plant species in National Plant Germplasm System

HDB	11	(Hybridoma Data Bank) immunoclones and monoclonal antibodies
HGIR	11	(Human Genome Information Resources) nucleotide probes under the Human Genome Initiative Project
HGMCR	11	(Human Genetic Mutant Cell Repository)
HGML	11	(Howard Hughes Medical Institute, Human Gene Mapping Library also called Yale Human Gene Mapping Library) human DNA probes and genetic maps
HIVSSA	4	(HIV Sequence and Analysis Database) AIDS virus amino acid and nucleotide sequences
IHIC	4	sequences of natural haemoglobin variants
ILDIS	11	(International Legume Database and Information Service) literature citations
JIPIDA	7	(Artificial Variant Database of the Japan International Protein Sequence Database) artificial protein variants, methods used, modification of the replacement
JIPIDB	7	(Biological Database of the Japan International Protein Sequence Database) protein properties and physical data
JIPIDM	6	(NMR Database of biopolymers of the JIPID) protein NMR data
JIPIDN	4	(Natural Variant Database of the JIPID) natural protein variants
JIPIDP	2,4	(The Japan International Protein and Nucleic acId Database of Physical properties) protein transition number, transition states, and conditions amino acid and nucleotide sequences
JIPIDS	3,4	(The Japan International Protein and Nucleic acId Database information from Asia and oceania) amino acid and nucleotide sequences
JIPIDSN	2,4	(Nucleotide Acid Sequence Database) nucleotide and amino acid sequences
JIPIDV	2,3,4	(Variant Database: International Protein Information Database) asia and oceania extension of JIPIDS,JIPIDP, and JIPIDB
LIMB	1	(Listing of Molecular Biology Databases) information on access to molecular biology databases
LIPDPHASE	11	(Lipid Phase Database) lipid properties and lipid phase behaviour
LYSIS	7	cleavage sites of endopeptidases
MBCRR	4	(MBCRR Protein Family Diagnostic Pattern Database and Search Tool) sequence patterns derived from homologue protein families in PIR
MEDLINE	11	(Database of the Medical Literature Access and Retrieval System) biomedical and molecular biology literature over 5 million citations and abstracts
MICIS	11	(Microbial Culture Information Service) data on the physical properties of cultured microorganisms from all UK National Culture Collections (over 30,000 strains)

MICROGERM	11	(Microbial Germplasm Database and Network) microbial germplasm news and information
MINE	11	(Microbial Information Network Europe: an Integrated Catalogue Network for a European Network of Microbial Culture Collection Databanks) microbial strain collection data
MIPS	4	(Martinsreid Institute for Protein Sequence data) Amino acid sequences
MOUSE	11	mouse DNA clones and probes and genetic maps
MOUSEMAN	11	linkage and synteny homologues between mouse and man
MSDN	11	(Microbial Strain Data Network) microbial strains and cultured cell lines
NAPRALERT	11	(NAPRALERT) literature references on isolation of chemical compounds from living organisms and/or pharmacology of natural products
NCYC	11	(UK National Collection of Yeast Cultures) yeast culture catalogue including probabilistic yeast identification system
NEWAT	4	(NEWAT) amino acid sequences
OLIGONIC	4	chemically synthesised oligonucleotide sequences
OMIM	11	(Online Mendelian Inheritance in Man) human inherited diseases (over 4500) loci and maps
PDB	6	(Protein Data Bank) crystallographic structure data for biological molecules
PIR	4	(Protein Identification Resource) amino acid sequences and bibliographic citations, sequences in preparation, and additional fragmentary and predicted sequences
PKCDD	4	(Protein Kinase Catalytic Domain Database) classification and alignment of new protein kinase
PMD	9	(Protein Mutant Database) protein mutants artificially derived
PPR	11	(Plasmid Prefix Registry) plasmid designation and names
PRCTR	11	(Plasmid Reference Center Transposon Registry) transposons
PRFLITDB	11	bibliographic literature on nucleic acids and peptides from over 1000 journals
PRF-SEQDB	8	(PRF-SEQDB) amino acid sequences
PROSITE	9	(PROSITE) biologically significant protein sequence patterns
PSEQIP	4	(PseqIP) nonredundant amino acid sequences
PSS	6,7	(Protein Secondary Structure database) crystal composition of proteins secondary structure
PTG	4	(Protein Translation of GenBank) amino acid sequences

QTDGPD	7	(Quest 2D Gel Protein Database) protein identification from 2D gel electrophoresis
RED	7	(Restriction Enzyme Database) restriction endonucleases more than 750 enzymes and their 130 different recognition sequences
RFLPD	11	(CEPH Public Database) polymorphic markers for high resolution map
RIBOSOMAL PROTEINS	11	sequence data of ribosomal proteins
SEQANALREF	11	(Sequence Analysis Literature Reference Data Bank) all references on sequence analysis
SIGPEP	4	(Signal Peptide Compilation) preprotein signal peptide sequence
SIGSCAN	3	(SIGnal SCAN) can be searched for eukaryotic transcription factor elements in DNA sequence
SRRSD	2	(16S Ribosomal RNA Sequence Database) nucleotide sequences of 16S ribosomes
SRSRSC	2	(Small Ribosomal Subunit RNA Sequence Compilation) small RNA Sequences
SVFORTYMUT	11	(SV40 Large T Antigen Mutant Database) SV40 large T antigen deletion and insertion mutants
SWISS-PROT	3,4	(SWISS-PROT Protein Sequence Data Bank) aligned amino acid sequences
TFD	various	(Transcription Factor Database) transcription factors and their cognate sequences
TRNAC	2	(tRNA Compilation) tRNA gene nucleotide sequence and alignment
VECBASE (VECTOR)	3	(cloning VECtor sequence dataBASE) cloning vector nucleotide sequences

Figure 7.2. Online databases for biological macromolecules and their their components (Source: Keil, 1990, p.54-56; Lawton et al., 1989; Colewell, 1989, p.351).

Type of database: 1, general information on databases; 2, RNA sequences; 3, DNA sequences; 4, amino acid sequences; 5, carbohdrate sequences; 6, macromolecular structure; 7, macromolecules,specific activity,other characteristics; 8, peptides; 9, amino acids; 10, crystallography,small molecules; 11, bibliographic and other.

Further information on these databases (including their addresses) can be obtained from:

von Heijne (1987, p.153)
Doolittle (1986, p.4)
Lesk (1988, p81, p.99)
Colewell (1989, p.351).
Ferguson (1989, p.24).
LIMB release 2.0 August 1990 (Burks and Keen, 1990)

7.4. THE MAIN SEQUENCE DATABASES

7.4.1. The GenBank database

The Genbank project is funded through IntelliGenetics Inc. (IG), which also supplies a suite of sequence analysis programs, and sponsorship is from the U.S. National Institute of Health, National Library of Medicine, Department of Energy and several other bodies (Burks *et al*., 1990). In 1989 it contained nearly 29,000 entries with 34.8 million DNA and RNA sequences and their description and bibliographic references (information from GenBank, Release 61, September, 1989). By August 1990 this figure was 35,000 sequences (Watts, 1990, p.40). Data are compiled by the Theoretical Biology and Biophysics Group at Los Alamos National Laboratory and distributed by several methods (as will be seen in the following):

(1) 9-track magnetic tape,

(2) floppy diskettes

(3) printed supplements to *Nucleic Acids Research*

(4) through the GenBank online service (GOS)

(5) through contract with the U.S. National Institute of Health GenBank System (Cuadra/Elsevier, 1990),

(6) through secondary distributors (e.g. IRL Press, Oxford, U.K.)

Data collection and distribution has been shared with the EMBL Data Library since 1982 and with the DNA Data Bank of Japan (DDBJ) since 1987. For a sample entry from GenBank see Burks *et al.*, (1990, p.7); Bilofsky *et al.*, (1988, p.45); or Claverie (1988, p.86).

7.4.2. The EMBL Data Library

The European Molecular Biology Laboratory (EMBL) in Heidelberg established the Data Library in 1980 to provide nucleic acid sequence data for research in Europe (Lesk, 1988). It has now a number of different databanks such as the EMBL Nucleotide Sequence Database with 27.2 million bases (February 1989) annotated and arranged in a similar way to GenBank entries (see sample in Kahn and Cameron, 1990, p.24). Recently it has been restructured using the ORACLE relational database management system to offer better and more efficient use of its content. The other change is that since May 1989 (Release 19) a separate "ancillary" database is in use for those new sequences which in part may already be in the database resulting from sequence entries merged to existing ones in GenBank which exchanges all its data with EMBL.

A database of restriction enzymes (RED) collected by Dr. R. Roberts (Cold Spring Harbor Laboratory) is available with all the releases of the nucleotide sequence data. The third databank is the SWISS-PROT (Protein Sequence Database) maintained by Amos Bairoch (University of Geneva). This is a collection of amino acid sequences from the Protein Identification Resource (PIR, Washington D.C.) and translated coding sequences of the EMBL Nucleotide Sequence Database. SWISS-PROT entries have pointers to related data in different databanks (PIR, EMBL Nucleotide Sequence Database and the Protein Data Bank at the Brookhaven National Laboratory).

In 1988 the EMBL started to distribute the Eukaryotic Promoter Database (EPD), a collection of references to sequences of eukaryotic transcription start sites from the Nucleotide Sequence Database prepared by Philip Butcher.

Access is by magnetic tapes (four releases a year), on CD-ROM or through online networks (GOS, BITNET/EARN, EMBNet or any other network such as JANET in the U.K. or ARPANET in the U.S.A. which has a gateway into one of these). In 1988 a European molecular biology network (EMBNet) was established to increase the use of the database with national centres such as SERC Laboratory (Daresbury, U.K.), COAS/CAMM Centre (Nijmegen, The Netherlands), Hoffman-La Roche (Basel, Switzerland) and CITIZ (Paris, France). For information on how to send electronic mail messages to a BITNET/EARN address see Kahn and Cameron (1990, p.30). For a sample record see Claverie (1988, p.87).

7.4.3. GENINFO

GENINFO was created as a 'static' database to contain data stored in other databases such as GenBank and EMBL databases. Most databases change haphazardly, with data being deleted and changed as well as being added. This causes all sorts of problems in that what seemed to be a 'fact' at one time quickly ceases to be true at a later time. The problem is becoming even greater now that revisions by one of the big three databases (GenBank, EMBL, and the DNA Database of Japan) will in future cause an automatic update in the other two. The intention is that GENINFO will keep data intact and unchangeable with entry dates so that users will know when data was recorded. (The problem of determining how many and what kind of revisions were undergone by a piece of data would thus be easilly solvable since the GENINFO database would have catalogued such a history of revisions). Thus the database is very different from all the others described here - it is static, and also it is much larger. It is growing at around 60,000 to 100,000 sequences per year, while GenBank only contains 35,000 sequences.

The database was set up by the U.S. National Library of Medicine. The unusual size and distinctive purpose of the database have created some interesting intellectual problems. The database must be able to accept information from all over the world without changing its basic structure too quickly, since rapid change in database structure would cause difficulties for the programmers preparing search routines.

7.4.4. The NBRF-PIR protein sequence database

The protein sequence database at the National Biomedical Reasearch Foundation (NBRF) has been maintained since the early 1960s to support research on evolutionary relationships between proteins. Later it was restructured into a fully integrated software-data system. The data is organised in terms of superfamilies, families, subfamilies, entries and subentries according to sequence similarity. This concept is likely to change in the near future and will be replaced by a system based on sequence alignment (i.e. each alignment will define a set of related sequences or subsequences). This will provide information about secondary and tertiary structure, function, active site or functional site location, evolution and chemical mechanism and will improve structure predictions further (Barker et al., 1990).

The database contains over 7,000 partial and whole protein sequences of about 2 million residues with bibliographic references and features of interest (e.g. binding site, modified site, disulphide bonds). The database format has been altered in line with CODATA recommendations with several advantages such as the alternative display of records using the PSQ program and records with more detailed information (for a sample entry see Barker et al., 1990, p.39).

Although most of the data is fully annotated (analysed and reviewed by scientists) the growth in the amount of sequence data has made it necessary to create a separate "new data" file with minimally reviewed entries to cope with the backlog of published data. This situation may change with recent improvements in data collection and methods in annotation. Another improvement is collaboration with other protein sequence databanks on an international level.

Distribution of the database is on magnetic tape and more recently on CD-ROM. It also can be searched online using the Protein Identification Resource (PIR) service or the GenBank Online Service (GOS) or through networks which have access to either of these.

For information contact MIPS, Martinsried Institut für Proteinsequenzen, Max-Plank Institut für Biochemie, D-8033 Martinsried bei Munchen, Germany or for electronic mail: MEWES@MIPS.BITNET is the code to use.

7.4.5. The NEWAT database

This contains essentially a subset (only 300,000 residues and 1250 sequences) of the PIR material although searches are in consequence faster. There are six sections in the database : enzymes from all sources, non-vertebrate eukaryotic sequences excluding enzymes, prokarytic sequences excluding enzymes, vertebrate sequences, animal virus sequences, and ribosomal sequences. The database is useful if one's search problem fits into the simple classes described here. The database is the work of Doolittle at the University of California at San Diego. However it has not been added to since 1985.

The NEWAT database demonstrates some of the problems that beset databases in molecular science. Doolittle was looking for the frequency of occurrence and relative position of amino acids and so he omitted from consideration near-duplicate sequences. He was interested to know if dipeptides or tripeptides occurred more frequently in proteins than would be expected by chance occurrence. So the database is far from complete or even from being representative. In fact in common with most databases it reflects the fact that it was created with a particular set of research questions in mind, which necessarilly has led to a limited and unrepresentative content. Another example of the specificity of motivation behind the creation of particular databases is the NBRF database mentioned above, which was set up to study protein evolution.

7.4.6. The PRF-SEQDB Databank

The Peptide Institute, Protein Research Foundation (PRF), an independent research institute for the synthesis of biologically active peptides in Japan. established a database of peptide and protein sequences (SEQDB), a relational database of the literature on peptides (LITDB) and another on synthetic compounds (SYNDB). In April 1988 the SEQDB had nearly 12,000 records with a sum of around 3 million amino acid residues (Seto _et al_., 1988). A retrieval program (Protein Information Analyser System or PRINAS) was developed and is distributed with the database on floppy diskettes.

The Peptide Institute started to collect papers related to research in 1962. It has been editing the secondary information journal "Peptide Information" since 1975. The data have been compiled in computer files and have become the Literature Database (LIT DB). The LIT database contains mostly citations but also some abstract information on 5000 references, and 20 indexes exist for keyword searches. The MAP database contains 1500 references to genes and is cross referenced to the LIT database. Each entry contains information similar to that found in the table of the Nomenclature and Chromosome Committee of the Human Gene Mapping Workshop. The PROBE database contains information on some 2000 probes or clones, cross-referenced to the LIT database.

7.4.7. The DNA Data Bank of Japan (DDBJ)

The nucleotide sequence databank of Japan (DDBJ) was established in 1983 with support from the Ministry of Education, Science and Welfare of Japan with cooperation of other bodies such as EMBL and the Los Alamos National Laboratory (Seto *et al.*, 1988, p.27). In 1985 it had 4.5 million bases. The PRINAS software is used for data retrieval and analysis.

7.4.8. GENESEQ

This patent database from Derwent became available in May 1990 (*Derwent Patents News* 1990). It contains nucleic acid and protein sequences taken from published patent applications (i.e. the information will be available in GENESEQ long before it is published in journals). It is the result of collaboration between Derwent and IntelliGenetics Inc. It is in a format suitable for searching by the IntelliGenetics (IG) suite of sequence analysis programs. The database contains all nucleotide sequences of a length greater than nine bases, all protein sequences greater than three amino acids in length, and probes of any length, which have been discovered in published patents. Currently, the database contains material from patents issued in 1989 only. Earlier patents will be added retrospectively. Besides material similar to that in other molecular databases such as substance name, author's name, keywords, method of synthesis, date, amino acid or nucleotide sequence, and so on, there is a substantial "claim" section which allows a short description of the claimed attributes of the sequence referred to in the patent.

7.4.9. Institut Pasteur Databanks

These comprise two databanks compiled by the Computer Science Group at Institut Pasteur, Unite d'Informatique Scientifique, Jean-Michel Claverie, Institut Pasteur, Paris, France. They are:

(1) PGtrans (or PTG) is a computer translation of the protein coding regions of GenBank,

(2) PseqIP is a non-redundant database constructed by merging the contents of several databanks.

7.5. DATA STRUCTURE AND MANAGEMENT

Databases contain the molecular sequences, plus bibliographic information. Information also exists on sequence annotation describing conflicts or uncertainties by different authors. The data structures used to store molecular sequences are usually linear (i.e. lists rather than trees or graphs). The assumption is that the user can convert this information into hierarchical, non-linear arrangements as he or she wishes at his or her laboratory workstation. Sequence data is annotated by means of look-up tables where the features of particular points in the sequence are noted.

Kanehisa comments on the current organisation of data in sequence databases:

> "There is an inherent problem in the way annotation is done. The current sequence databases are organised in such a way as to enter a sequence first and then add whatever information is associated with this sequence. This is a kind of top-down approach... An alternative

> is a bottom-up approach, namely starting from an interest in one biological function site, all sequences representing this function are collected. This approach has been taken by a number of individual researchers. There are established collections of splicing sites and promoter sites in the nucleic acid sequence and attempts to collect so-called functional motifs, sometimes also called templates and fingers, in the amino acid sequence. These collections tend to be well verified because they are made by specialists in the field. Although annotation in the sequence databases was originally designed to generate this kind of collection, the current features table approach is neither flexible nor simple enough to meet biological complexity" (Kanehisa, 1989, p.229).

7.5.1. Management of databases

This involves two activities.

(1) There is data acquisition, involving extraction of published data and formatting it into acceptable arrangements. Data extraction is expensive and time consuming. Often the important details are buried deep in the original text, perhaps existing implicitly and discoverable only by trained personnel (Seto *et al.*, 1989).

(2) Also there is distribution and marketing of the database. Distribution involves not only the making available of data but also of a user manual and a number of indexes. Sponsorship has remained uncentralised, and for this and other reasons (see Keil, 1990, p.53) the database system has continued to spread.

Distribution is undergoing some change as the existing organisations collaborate over the setting of conventions and standards of content and format. At present each databank has its own format but these are similar in structure, including as they all do a documentary part and sequence, but differing in the number of fields and in field iden-

```
ENTRY           OKBOG         Protein  #Length 670 #Checksum 5530
NAME            cGMP-dependent protein kinase (EC 2.7.1.37) - Bovine
DATE            17-May-1985  #Sequence 17-May-1985  #Text 27-Nov-1985
SPECIES         Bos taurus #Common-name ox
REFERENCE       Sequences of residues      1-17, 89-374, and 407-670
   #Authors     Takio K., Wade R.D., Smith S.B., Krebs E.B., Walsh, K.A.
                  Titani, K.
   #Journal     Biochemistry (1984) 23: 4207-4218
REFERENCE       Sequence of residues 13-104
   #Authors     Takio, K., Smith, S.B., Walsh, K.A., Krebs, E.G., Titani, K.
   #Journal     J. Biol. Chem. (1983) 258:5531-5536
REFERENCE       Sequence of residues 373-409
   #Authors     Hashimoto, E., Takio, K., Krebs, E.G.
   #Journal     J. Biol. Chem. (1982) 257: 727-733
COMMENT         The protein, isolated from lung, is a dimer of identical chains.
SUPERFAMILY
   #Name        cAMP-dependent protein kinase regulatory chain
                  #Residues 102-340
   #Name        kinase-related transforming protein
                  #Residues 475-599
KEYWORDS        acetylation\ phosphoprotein\ cGMP\
                  serine-specific protein kinase
FEATURE
    1                     #Modified-site acetylated amino end\
    42                    #Disulfide-bonds interchain\
    58                    #Binding-site phosphate\
    1-101                 #Domain dimerization <DIM>\
    102-219               #Domain cGMP-binding 1 <GB1>\
    320-340               #Domain cGMP-binding 2 <GB2>\
    341-474               #Domain ATP-binding <APB>\
    475-599               #Domain catalytic <CAT>
COMMENT         These boundaries are approximate.
SUMMARY         #Molecular-weight 76287 #Length 670 #Checksum 5530
SEQUENCE
                5         10        15        20        25        30
      1 S E L E E D F A K I L M L K E E R I K E L E K R L S E K E E
     31 E I Q E L K R K L H K C Q S V L P V P S T H I G P R T T R A
     61 Q G I S A E P Q T Y R S F H D L R Q A F R K F T K S E R S K
     91 D L I K E A I L D N D F M K N L E L S Q I Q E I V D C M Y P
    121 V E Y G K D S C I I K E G D V G S L V Y V M E D G K V E V T
    151 K E G V K L C T M G P G K V F G E L A I L Y N C T R T A T V
    181 K T L V N V K L W A I D R Q C F Q T I M M R T G L I K H T E
    211 Y M E F L K S V P T F Q S L P E E I L S K L A D V L E E T H
    241 Y E N G E Y I I R Q G A R G D T F F I I S K G K V N V T R E
    271 D S P N E D P V F L R T L G K G D W F G E K A L Q G E D V R
    301 T A N V I A A E A V T C L V I D R D S F K H L I G G L D D V
    331 S N K A Y E D A E A K A K Y E A E A A F F A N L K L S D F N
    361 I I D T L G V G G F G R V E L V Q L K S E E S K T F A M K I
    391 L K K R H I V D T R Q Q E H I R S E K Q I M Q G A H S D F I
    421 V R L Y R T F K D S K Y L Y M L M E A C L G G E L W T I L R
    451 D R G S F E D S T T R F Y T A C V V E A F A Y L H S K G I I
    481 Y R D L K P E N L I L D H R G Y A K L V D F G F A K K I G F
    511 G K K T W T F C G T P E Y V A P E I I L N K G H D I S A D Y
    541 W S L G I L M Y E L L T G S P P F S G P D P M K T Y N I I L
    571 R G I D M I E F P K K I A K N A A N L I K K L C R D N P S E
    601 R L G N L K N G V K D I Q K H K W F E G F N W E G L R K G T
    631 L T P P I I P S V A S P T D T S N F D S F P E D N D E P P P
    661 D D N S G W D I D F
//
```

Figure 7.3. An example of CODATA recommended format in the NBRF-PIR database. (Source: Bishop et al., 1987, p.96).

tifier. For example, the word "ACCESSION" is used in GenBank and the two-letter code "AC" is used in EMBL entries even though they mean the same thing and even though the databanks exchange data. Other important fields are "FEATURES" and "SITES" in GenBank and "FH" and "FT" in EMBL, which hold information on coding regions, start and stop signals for transcription and translation, mutations, conflicting experimental results on the sequence, and so on. The same relationship exists between entries in the PRF-SEQDB and PGtrans entries (see formats in Claverie, 1988). Some general principles were set down after a meeting in 1984 of the CODATA Task Group on Coordination of Protein Sequence Data Banks. These recommendations have been put into effect by the NBRF-PIR database (George et al., 1987a, 1987b). Figure 7.3 shows an example of CODATA recommended format in the NBRF-PIR database. Most of the information in Figure 7.3 is self-explanatory. The information in the "FEATURE" section provides comments at positions of interest in the sequence.

7.5.2. Cooperation between databanks

A new development is the cooperative projects which are beginning to take shape.

One example of cooperation is that by the Protein Identification Resource (PIR) in Washington which has made cooperative agreements with similar organisations in Germany and Japan. This has involved the establishment of PIR-International, an association of protein sequence data collections including NBRF, the Martinsried Institute for Protein Sequences (MIPS) in West Germany, and the International Protein Information Database in Japan (JIPID). Each organisation agrees to collect data from its own region and then to share them. This development has been motivated by the realisation that "the time has long since passed when one database could adequately collect all the sequence data being determined and still maintain the quality of input, frequent updating and efficient publication of ever increasing data files..." (Tsugita, 1989, p.91). The integrated database consists of a number of protein and nucleic acid sequence databases and software designed for sequence analysis (Barker et al., 1990). The databases include the NBRF Protein Sequence Database and a database of protein sequences in preparation for inclusion in the NBRF database; a protein database translated from the nucleotide sequences in GenBank (R) by J.M. Claverie of the Pasteur Institute; the NBRF Nucleic Acid Database; and a reformulated version of the European Molecular Biology Laboratory's (EMBL) Nucleotide Data Library. This information is now available in the form of the following databases:

(1) the Biological Activity Database (JIPIDB), containing data on the biological and chemical activity of proteins, and data files for enzymes, toxins, and electron-carrier proteins,

(2) the Variant Database (JIPIDA), containing information about artificially created mutant molecules, and

(3) the NMR database (JIPIDM), containing literature and also coordinate data.

This collaboration has resulted in the publication of the journal Protein Sequences and Data Analysis which aims to publicise recent database entries.

The second example of cooperation is that by the EMBL Data Library, set up in 1980, which began international collaboration in 1982. It collaborates with GenBank and the DNA Database of Japan. Each of the three members of this collaboration collects a portion of the reported sequence data and exchanges it with the others on a regular basis.

7.5.3. Data acquisition

Although databanks would like to be the first and primary recipients of sequence data, so that data are electronically mailed to them, professional career priorities necessitate that scientists publish first and afterwards send the data to the database , or as is more usual, fail to send it, leaving it up to the database to seek and extract data from the journal in a painfully slow process (Doolittle 1990). In a survey in 1984, 95% of new sequences were derived from the top 30 journals and 5% from another set of 52 journals (Seto et al., 1988, p.31).

Databases sometimes miss published material, or they sometimes contain duplicate items. Removal of duplication involves merging of one protein sequence with a protein subsequence (for example, the smaller adenovirus items with genomic items). However, this produces very long entries which are difficult to annotate. This problem is dealt with by splitting entries into parts. However, splitting often disrupts biologically significant properties.

7.6. DATA RETRIEVAL AND MANIPULATION

7.6.1. Search strategy

Although databanks such as GenBank and EMBL Data Library containing nucleic acids (i.e. DNA sequences) are four times larger than those such as NBRF-PIR which contains amino acid sequences (i.e protein sequences), there is a lot of sense in searching the latter type of database first.

> "Currently the biggest sequence data banks, GenBank and the EMBL Data Library, are mostly collections of DNA sequences. Nonetheless, most of the interesting matches that have been made in the last decade have involved protein sequences. Why is this so, and how does it bear on the maintenance of sequence data banks? Certainly, the overwhelming majority of new "protein sequences" are being determined on the basis

of DNA sequences, so it is altogether appropriate and indeed desirable to bank them as the DNA sequences. The DNA sequence provides essential information, in the resource sense, for molecular geneticists and others; there is an enormous amount of molecular biology not directly concerned with the gene product per se. Nonetheless, there are good reasons for searching expeditions to begin at the protein sequence level.

Some investigators are under the erroneous impression that there is more to be gained by searching the actual DNA sequence rather than the amino acid sequence derived from it. That view is greatly mistaken, and resemblances will be missed if it is adopted. The reason is, of course, that there are only four bases but 20 amino acids. As such, the so-called signal-to-noise ratio is improved greatly when the DNA sequence is translated; the "wrong-frame" information is set aside and third base degeneracies consolidated. As a general rule, then, searchers dealing with potential gene products should translate their DNA sequences into the protein equivalents. Obviously, this implies that the sequence bank to be searched should be in the form of protein sequences also" (Doolittle, 1990, p.99).

7.6.2. Database management systems

Data entries in molecular sequence databases vary considerably in size. For this reason there has not been a widespread takeup of general purpose database management systems. Rather, specialised programs have tended to be used for accessing particular types of information for particular purposes. These specialised programs are usually integrated into less specialised data manipulation and analysis programs. A number of basic operations are available on most retrieval systems. The way in which they are implemented depends on a number of factors:

(1) The way the data is arranged can vary considerably. Some simple systems put each entry into a different file. The search process then depends upon the filename and the complexity of the operating system. For example, MSDOS and UNIX allow wildcard or "*" to mean any string of any characters, so that "find prot*" finds files "protein" and "protase" if they exist. Most systems have several entries in the same file. For example each BIONET file contains several entries belonging to the same organism or

species. The most complex method of data organisation involves use of database management systems. These have their own data definition language (DDL) and data manipulation language (DML). The trend in recent years has been to relational databases, where the user sees only tables, rather than having to deal with programming details such as arrays, indexes, and pointers. Another important modern standard is for multi-user systems, which allow simultaneous access for many different users. Remote access capabilities are increasingly in demand. Additionally, data-independence is important: when the format of files is changed (eg a new record, or a new field) the system should be such that the data entry and manipulation programs do not need to be changed. It is important to have data control, or the automatic elimination of duplicate files. A related requirement of modern systems is that each file should have only one machine address, and each user one identifier or password. All these requirements are met in modern database management systems. These advantages over simpler systems are paid for by the greater cost and the longer time needed for the user to learn the associated query language.

(2) The way the data is retrieved also varies. One important facility available in many systems is retrieval functions which use look-up tables to enable parts of a sequence to be found. These functions are then used to build up consensus patterns (i.e. parts of sequences common to two or more proteins). The consensus patterns are then used to gain information about an unknown protein from another protein whose identifier is known but which shares a common subsequence with the unknown protein. The means of retrieval can vary widely. When the sequence identifier (or name) is known

then the sequence can be retrieved just by stating the identifier required. Sequences can also be retrieved by inputting any subsequence. For example the sequence a1...a100 can be found by inputting a5...a10, for a=amino acid. First a1...a5 from the sequence is pairwise compared with amino acids in the subsequence, then a2...a6, then a3...a7, and so on. This is essentially what the QUEST program does in the BIONET facility. Sequences can also be identified by keyword. This method relies on informative and accurate annotations since it is these that provide the information selected by the keyword. An example of the keyword method is provided by the PSQ program which is available with the PIR database. The most important type of retrieval method is retrieval by sequence similarity. This means that search is not for completely identical sequences, but ones which are similar. As organisms evolve and become more complex, particular amino acids become supplemented by sub-sequences of amino acids. The original amino acid remains but its biological function is lost so that it should be omitted from consideration when searching for similar sequences. Similar, non-identical sequences have important evolutionary relationships.

(3) Molecular sequence databases depend upon data management and manipulation software which is common to other molecular information. However, there are some types of problem which are particular to these types of database. For example, codon usage patterns are important indicators of sequence similarity and look-up tables and search procedures of them are useful. An example is the NAQ program, which can work with both protein sequence and nucleic acid databases. Another type of analysis is the classification of known sequences into superfamilies using cluster, fac-

tor or discriminant analysis. Cluster analysis attempts to create groups for which within-group variance is minimised and for which between-group variance is maximised. Factor analysis is useful for taking a number of items evaluated in a number of attributes and saying what the "major" attributes are, and so is useful for reducing massive amounts of data to manageable proportions. Discriminant analysis takes a known item (such as a sequence whose identity is known) and says which of a number of other items is the "nearest" one to it. Another important problem is the predicting of regions which have biological significance. For example, the SASIP system can scan databases and determine the frequency of occurrence of say, octanucleotides (from the GenBank database) and tripeptides (from the PIR database). Information theory can be used to predict that the least frequently occurring oligomers carry the greatest amount of information and are thus most likely to be biologically important.

7.7. AN EXAMPLE OF PROTEIN SEQUENCE ANALYSIS SOFTWARE

An example of how specialised protein sequence manipulation functions are often blended with more standard ones for database management is provided by the PRINAS software used for managing the PRF-SEQDB database. PRINAS consists of amino acid sequence database, sequence analysis functions, and standard database management functions (Seto *et al.*, 1988). The major functions provided are:

(1) Keyboard or from-file input of DNA base sequence data with screen help and error checking.

(2) Automatic translation to amino acid sequence data and output to screen or printer.

(3) Computation of amino acid composition and molecular weight and partial specific volume.

(4) Dot matrix representation of sequence homology between two amino acid sequences. Output for interpretation by user.

(5) Prediction of secondary structure from the amino acid sequence.

(6) Calculation of hydrophobicity and hydrophilicity character.

(7) Prediction of the peptide type.

(8) Keyword search of database for peptides containing a particular amino acid sequence exactly.

(9) Search of database for peptides containing fragments of the particular amino acid sequence.

(10) Search of the database for homologous sequences, representing peptides containing amino acid subsequences which are homologous to the sequence whose identity is unknown.

The last step can be illustrated by an example, which shows that the only two homologues of the amino acid sequence HKQGPANLGLF are :

```
      HKVGP-NLWGLF
with
      HKQGPANL-GLF     (target sequence )


and
      HKTGP-NLHGLF
with
      HKQGPANL-GLF     ( target sequence )
```

In order to arrive at a homology the target sequence HKQGPANLGLF had to be given an inserted gap (between the L and the G) in both cases. The homologues corresponded over the sequence HK.GP.NL.GLF, where dot means non-correspondence. So both homologues were successful on 9 out of the 12 positions in the sequence.

7.8. REFERENCES

Ash J.E., Chubb P.A., Ward S.E., Welford S.M., Willett P., 1985, Communication storage and retrieval of chemical information, Ellis Horwood, Chichester.

Barker W.C., George D.G.,and Hunt L.T., 1990, Protein sequence database, 31-49 in Doolittle R.F., Molecular evolution: computer analysis of protein and nucleic acid sequences, Methods in Enzymology,183, Academic Press, New York.

Bilofsky H.S., Burks C., Fickett J.W., Goad W.B., Lewitter F.I., Rindone W.P., Swindell C.D., and Tung C-S., 1988, GenBank 43-54 in Lesk A. M., (Ed.), Computational molecular biology: sources and methods for sequence analysis, Oxford University Press, Oxford.

Bishop M.J., Ginsburg M., Rawlings C.J., and Wakeford R., 1987, Molecular sequence databases, 83-114 in Bishop M.J., and Rawlings C.J., (Eds.), Nucleic acid and protein sequence analysis: a practical approach, I.R.L. Press.

Burks C., Cinkovsky M.J., Gilna P., Hayden J.E.-D., Abe Y., Atencio E.J., Barnhouse S., Benton D., Buenafe C.A., Cumella K.E., Davison D.B., Emmert D.B., Faulkner M.J., Fickett J.W., Fischer W.M., Good M., Horne D.A., Houghton F.K., Kelkar P.M., Kelley T.A., King M.A., Langan B.J., Lauer J.T., Lopez N., Lynch J., Marchi J.B., Marr T.G., Martinez F.A., McLeod M.J., Medvick P.A., Mishra S.K., Moore J., Munk C.A., Mondragon S.M., Nasseri K.K., Nelson D., Nelson W., Nguyen T., Reiss G., Rice J., Ryalls J., Salazar M.D., Stelts S.R., Trujillo B.J., Tomlinson L.J., Weiner M.G., Welch F.J., Wiig S.E., Yudin K., and Zins L.B., 1990, GenBank: current status and future directions, 3-23 in Doolittle R.F.,

(Ed.), Molecular evolution: computer analysis of protein and nucleic acid sequences, Methods in Enzymology, 183, Academic Press, New York.

Burks C. and Keen G., 1990, LIMB DATABASE: Listing of molecular biology databases, Release 2.0, Los Alamos Laboratory,New Mexico, USA.

Claverie J.-M., 1988, Computer access to sequence databanks 85-99 in Lesk A. M., (Ed.), Computational molecular biology: sources and methods for sequence analysis, Oxford University Press, Oxford.

Colewell R.R., 1989, Biomolecular data : a resource in transition, Oxford Science Publications, Oxford.

Cuadra/Elsevier, 1990, Directory of online databases, Elsevier.

Dayhoff M.O., (Ed), 1965, Atlas of protein sequence and structure, 5,Natl. Biomed. Res. Found.,Washington,DC.

Derwent Patents News, 1990, "GENESEQ: a revolutionary new database in biotechnology",May.

Doolittle R.F., 1986, Of URFS and ORFS: A primer on how to analyse derived amino acid sequences, University Science Books, Mill Valley, California.

Doolittle R.F., 1990, Searching through sequence databases, 99-111 in Doolittle R.F., Molecular evolution: computer analysis of protein and nucleic acid sequences, Methods in Enzymology, 183, Academic Press,New York.

Ferguson J.J., 1989, "A directory of information resources related to biotechnology", Chemical Information Bulletin, April.

George D.G., Mewes H.W., and Kihara H., 1987a, "A standardised format for sequence data exchange", Prot.Seq.Data Anal., 1, 27-39.

George D.G., Barker W.C., and Hunt L.T., 1987b, The Protein Identification Resource (PIR):An online computer system for the characterisation of proteins based on comparisons with previously characterised protein sequences, 445-453 in L'Italien J.J., (Ed.), PROTEINS: structure and function, Selected proceedings of the first symposium of American protein scientists on modern methods in protein chemistry held September 30-October 3, 1985, San Diego, California, Plenum Press, New York.

George D.G., Hunt L.T., and Barker W.C., 1988, The National Biomedical Research Foundation protein sequence database, 17-26 in Lesk A.M., (Ed.), Computational molecular biology: sources and methods for sequence analysis, Oxford University Press, Oxford.

Holley R.W., Apgar J., Everett G.A., Madison J.T., Merrill S.H., Penswick J.R., and Zamir A., 1965, "Structure of a ribonucleic acid", Science, 147, 1462-1465.

Kahn P., and Cameron G., 1990, EMBL Data Library, 23-31 in Doolittle R.F., (Ed.), "Molecular evolution : computer analysis of protein and nucleic acid sequences," Methods in Enzymology, 183, Academic Press, New York.

Kanehisa M., 1989, Databases: what's there and what's needed, 227-236 in Colwell R.R., Biomolecular data: a resource in transition, Oxford University Press, Oxford.

Keil B., 1990, Cooperation between databases and scientific community 50-60 in Doolittle R.F., (Ed.), "Molecular evolution : computer analysis of protein and nucleic acid sequences," Methods in Enzymology, 183, Academic Press, New York.

Lawton J.R., Martinez F.A., and Burks C., 1989, "Overview of the LIMB database", Nucleic Acids Research, 17, 5885-5899.

Lesk A.M., 1988, The EMBL Data Library, 55-65 in Lesk A. M., (Ed.), Computational molecular biology: sources and methods for sequence analysis, Oxford University Press, Oxford.

Maxam A.M.,and Gilbert W., 1977, "A new method for sequencing DNA", Proc. Natl. Acad. Sci. USA., 74, 560-564.

Maxam A.M.,and Gilbert W., 1980, "Sequencing end-labelled DNA with base-specific chemical cleavages", Meth.Enzymol., 65, 499-559.

Sanger F., Nicklen S., and Coulson A.R., 1977, "DNA sequencing with chain-terminating inhibitors", Proc. Natl. Acad. Sci U.S.A., 74, 5463-5467.

Schroeder W.A., (Ed.),1968, The primary structure of proteins: principles and practices for the determination of amino acid sequnce, Harper and Row, New York.

Seto Y.,Ihara S., Kohtsuki S., Ooi T., and Sakakibara S., 1988, Peptide and protein databanks in Japan, 27-37 in Lesk A.M.,(Ed.) Computational molecular biology: sources and methods for sequence analysis, Oxford University Press, Oxford.

Seto Y.,Ihara S., Kohtsuki S., Ooi T., and Sakakibara S., 1989, Problems in maintaining a protein sequence database, 71-76 in Colewell R.R., Biomolecular data: a resource in transition, Oxford Science Publications, Oxford.

Tsugita A., 1989, The database crisis: an emerging Japanese database's problems and solutions, 91-95 in Colewell R.R., Biomolecular data: a resource in transition, Oxford Science Publications, Oxford.

von Heijne G., 1987, Sequence analysis in molecular biology: treasure trove or trivial pursuit, Academic Press, San Diego.

Watts S., 1990, "Making sense of the genome's secrets", New Scientist, 37-41, 4 August.

Chapter 8:

The Main Structure Databanks in Molecular Science

8.1. THE SIGNIFICANCE AND HISTORY OF STRUCTURE DATABANKS

Structure databanks contain atomic coordinates, bibliographic citations, and other information related to structural studies. Numeric structural data for organics, organometallics and metal complexes studied by X-ray and neutron diffraction are available from the Cambridge Crystallographic Data Centre (CCDC) and from other similar databanks worldwide (Allen et al., 1979, Allen et. al., 1987). This and the Inorganic Crystal Structure Database (CRYSTIN) in Karlsruhe, for example, may become essential for studying interactions of drugs with macromolecules or the coordination of ligands in metalloproteins. The Protein Data Bank (PDB) at Brookhaven National Laboratory (Bernstein et al, 1977) was established in 1971 as a computer-based system of three-dimensional structures of biological macromolecules and is used mainly to study protein structure and modelling (Koetzle et al., 1989). There are also a NRC Metals Crystallographic Data File (CRYSTMET) which contains 23,000 metallic phases, a Powder Diffraction File, and a U.S. National Bureau of Standards Crystal Data File (CRYSTDAT) which contains data on all crystalline solids for which the basic cell parameters are known: this amounts to about 130,000 compounds (Abola et al., 1988). Recently a carbohydrate structure database (CCSD) was established (U.S.D.O.E., 1986). with the "growing need to attach carbohydrate chains possessing particular attributes to engineered proteins" (Doubet et al., 1989).

There is a need to integrate protein and nucleotide sequences with structures as they get reported (Askigg et al, 1988) and cross-reference databanks with others. For example records in the carbohydrate databank are cross-referenced with Chemical Abstracts, patent literature, protein databases, and spectroscopic databases.

Recent developments in the protein structure databank, a detailed description of the carbohydrate structure databank, and some integrated databanks of structures and sequences, will be the subject of this chapter.

8.2. THE PROTEIN DATA BANK (PDB) AT BROOKHAVEN

While sequence databases have mainly been regarded as a source of homology searches, the Protein Data Bank (PDB, also called CRYSTPRO) at Brookhaven has had a big impact on stimulating theoretical studies toward understanding protein folding and structure. The PDB has over 400 sets of atomic coordinates of proteins although the number of structures determined to sufficient resolution that an atomic model can be generated is about 500 and the number of biological macromolecules that have been crystallised is about 1200. While many journals now encourage or require deposition of coordinates data the databank includes 106 bibliographic entries on structures for which coordinates have not yet been deposited. Most of the data are on enzymes and non-enzyme proteins (including viruses), and there are also some tRNA and DNA items (see table 7.2 in Abola et al., 1988). Data submitted to the databank are processed and entered into the standard format (Protein Data Bank, 1985). This includes IUPAC-IUB conventional atom and residue names, essential bond connectivity information, and citations. The databank requires that data should be in machine-readable form to reduce errors in atomic coordinate entries which contain three-dimensional coordinates for more than 1000 atoms at a time. It should be stressed that PDB differs from many other databanks because it stores structure information which is not published in the primary literature. It maintains a strong connection with scientific journals that require data deposit prior to or simultaneously with publication. The Commission on Biological Macromolecules of the International Union of Crystallography (C.B.M.I.U.C) introduced a policy on the verification of results from the early stages of analysis and on the publication and the deposition of data from crystallographic studies of biological macromolecules. This policy was recently published (C.B.M.I.U.C., 1989). The policy gives details on the level of description in the publication and the atomic coordinates and structure-factor information that should be deposited in the Protein Data Bank. It requires inclusion in the publication the statement that the atomic coordinates and structure-factor data described in the paper have been deposited in the Protein Data Bank at Brookhaven, giving its number as a reference. It also requires that users of deposited data should cite the primary references as well as the Protein Data Bank when making use of the data. Similarly some data (see Wlodawer et al., 1989) have also been deposited with the British Library Document Supply Centre. The distribution of the databank is on magnetic tape. The complete coordinate tape (DATAPRTP) contains all the coordinate files, bibliographic files on solved structures for which the coordinates were not deposited and several programs that utilize this information. The PDB does not supply any database management system. Users extract individual files for analysis (listed on microfiche or in the quarterly issues of the PDB newsletters). Also users can implement their own system according to their interest. Leaflets from commercial firms offering graphics systems that are compatible with the PDB data format and articles in the Journal of Molecular Graphics are the main sources of relevant information (Abola et al., 1988). The document Sources of Visual Aids for Molecular Structure can be obtained from Brookhaven. This document contains the names and addresses of those companies mentioned above.

This situation is likely to change due to the need for efficient searching facilities to study protein structures. The result of this effort is shown, for example, in the application of geometric searching algorithms that have been used for substructure searching in the PDB files for small three-dimensional molecules to search protein structures (Brint et al., 1989) or techniques derived from graph theory to compare secondary structure motifs in proteins (Mitchell et al., 1989). Mitchell et al., have devised a data representation method for storing partial information (just secondary structure for pattern searching) in such a way that search algorithms to discover similar types of structural patterns can work efficiently. The data representation involves storing helices and strands as graph nodes and inter-line angles and distances as edges. Other approaches, such as the application of a "two-path recursive relational database structure" are under development.

8.3. THE CAMBRIDGE CRYSTALLOGRAPHIC DATABANK (CRYSTOR)

The Cambridge Crystallographic Data Centre (CCDC) files comprise bibliographic, numeric and chemical connectivity data for organics, organometallics, and metals researched using diffraction and X-ray studies. It is the largest structural database, holding 74,000 items in 1989. The average number of atoms per structure has grown from 13 in 1960 to 37 in 1978, when there were only 25,000 items (Allen et al.,

1979). Figure 8.1 contains a summary of information in the CCDC database.

Bibliographic File (BIB)
Compound name; synonym (or trivial) name; qualifying phrase(s) indicating, e.g., neutron study, low temperature work, absolute configuration determined; molecular formula; author list; literature reference; cross-references to MSD; chemical class assignment(s).

Chemical Connectivity File (CONN)
Compact coded representation of the chemical structural diagram in terms of atom and bond properties for each residue in the crystal chemical unit.

Structural Data File (DATA)
Unit-cell parameters; space group; symmetry; atomic coordinates of crystal chemical unit; published bond lengths; accuracy indicators (R factors); evaluation flags; textual comment relating to errors located; corrections applied and details of disorder.

Figure 8.1. Summary of information contained in CCDC database. (Source: Allen et al., 1979).
Reprinted with permission from Database, Vol. 121, Allen F.C., and Ferrell W.R., "Numeric databases in science and technology: an overview", Copyright (c) 1989, Online Inc.

8.4. THE COMPLEX CARBOHYDRATE STRUCTURE DATABASE (CCSD)

In 1986 the number of published carbohydrate structures was between 6,000 and 9,000. These were scattered around in the literature and in private databases, the largest of which was of 650 items. This estimated number of published structures is growing with 1,000 new ones each year due to improved analytical techniques and greater scientific interest in their biological functions in intra- and extra-cellular communications, regulation, and immunological response. As an example of a very specific (though important) user need and of the kinds of research questions underlying the demand for information, a workshop was held on the specification and implementation of the new database (U.S.D.O.E., 1986).

It was suggested in that report that along with the nucleic acid and protein sequence databanks this new carbohydrate databank should have the following characteristics to serve the needs of scientists working in molecular science:

(1) to help to find structures which have been reported in the literature so far but which are not indexed even by Chemical Abstracts;

(2) to help to understand new kinds of carbohydrate structures "like an N-linked or O-linked glycoprotein or glycosphingolipid" (U.S.D.O.E., 1986, p.4);

(3) to check novelty of a structure and whether the authors have cited the literature relevant to their article;

(4) to compare carbohydrate structures like polysaccharides of different origin; it was found that "the number of bacterial polysaccharides with known structures is so high now that it's difficult or even impossible to memorize them".

A group of scientists working with plant cell-wall polysaccharides was interested in the research of groups which are studying other types of complex carbohydrates. According to one of these scientists

> "we are concerned with finding out the structures they are characterizing and with evaluating the techniques they are using to solve problems in structural analysis. I am personally getting swamped with the literature. I can't keep up with all the structures people are publishing, and I can only see the problem getting worse" (U.S.D.O.E., 1986, p.5).

Such a new carbohydrate database, it was argued, would assist the rapid development of molecular science in a number of ways.

(1) When seeking alternative targets for active sites of enzymes for example, a computerised database could allow searches for analogous glycosyl sequences or substructures in carbohydrates based on similar three-dimensional shapes (generated by computer modelling rather than just glycosyl sequence).

(2) It would be easier to compare known and newly defined structures and analyse them for evolutionary relatedness or compare them between tissues, organisms, and species or between diseased and normal cells.

(3) The use of computer calculations and nuclear magnetic resonance spectroscopy data gives extra value with three-dimensional information when, for example, a synthetic chemist wants to synthesise analogues of biologically active carbohydrates that can be used as drugs or pesticides.

(4) Interfacing the carbohydrate database with the existing protein database it would be easier to visualise glycoprotein molecules for further study.

> "Genbank can be cross-referenced for information that is related to nucleic acid sequences of different glycoproteins whose coding sequence is represented there" (U.S.D.O.E., 1986, p.7).

In 1989 the database contained 2000 entries and it is expected that by 1991 the number of entries will be 5,000 (Doubet et al., 1989). Each record in the database contains data on full primary structure of the complex carbohydrate, the citation, and supplementary text information such as keywords, biological activity, and so on. For an example of a record from the database see Doubet et al., (1989).The record fields include a chemical structure diagram, the CarbBank accession number, the date, author of citation, title of citation, citation source, determination, spectroscopic methods used in structure, biological source and activity, information about binding studies, key words, glycoconjugate information, cross-rcferences to other CCSD records, to Chemical Abstracts, to patent literature, to protein databases, to spectroscopic database, and to other databases.

The database contains many complex structures such as branched structures having a variety of glycosyl residues possessing non-glycosyl substituents, or containing a variety of points of attachment of glycosyl linkages, anomeric configurations, and ring forms. These require special software for searching and one such is CarbBank (Doubet et al., 1989, p.475) which can search 2000 records in less than 2 seconds on a modern PC. Bearing in mind the complexity of each structure, this is impressive.

Only 75 percent of the 2000 entries in 1989 were unique owing to the fact that many structures were assumed to be new when they were not. This problem of the same structure being "discovered" simultaneously, more or less, by independent researchers, will be reduced as the database comes into more widespread use.

The entries have been provided by some 26 curators expert in such areas as N- and O-linked carbohydrate chains of glycoproteins,glycolipids, and bacterial and plant polysaccharides.

The difficulties of carbohydrate structure prediction and a computer program for structure analysis of polysaccharides (sugar molecules of repeating units) called CASPER are described by Jansson and others from the Department of Organic Chemistry, Arrhenius Laboratory, University of Stockholm (Jansson et al., 1989). Similar approaches (using sugar and methylation analysis data in combination with unassigned 13C-n.m.r. chemical shift data) have been reported for the analysis of linear bacterial polysaccharides and mannans (see references 9 and 10 in Jansson *et al*., 1989, p.91)

8.5. INTEGRATED ACCESS TO STRUCTURE AND SEQUENCE DATA

Although in the past there has tended to be separate work on one-dimensional sequence searching and three-dimensional structure searching, molecular biology is developing in such a way that it is becoming necessary to be able to access these two types of data simultaneously. Enzyme analysis, antibody-antigen interactions, and receptor-ligand interactions are good examples of the need to understand three-dimensional structure before it is possible to understand protein function at the molecular level. Also effective and systematic design of modified proteins, ligands and even novel molecules is needed at the three-dimensional coordinate level.

One example of the need for simultaneous access is in the problem of alignment of distantly-related sequences. Using amino acid sequences by themselves, it is not possible to align the sequences of distantly-related proteins. In particular, for those protein pairs which have sequence homologies between 15 and 20 percent residue identity, the Needleman-Wunsch-Sellers algorithm gives different results from those derived from three-dimensional structures. One type of error which often appears in sequence-based alignments of distantly-related proteins is the introduction of gaps in the interiors of helical regions. When protein structures are investigated, it is found that such gaps do not occur, because such gaps are inconsistent with the force constraints necessary for protein structure stability. Lesk (1989, p.190) comments:

> "Specifically, an insertion or deletion of one or more residues in the middle of a helix would entirely change the pattern of residue-residue packing at helix-helix interfaces and would destroy the topographic complementarity of the occluding surfaces".

This observation has been made for a number of different proteins and has led to successful corrective algorithms. Another research problem for which simultaneous access to sequence and structure databases would be useful is the discovery of the as yet unknown way in which the amino acid sequence determines its three-dimensional structure. This is one of the most significant current unsolved puzzles in molecular biology. Some encouraging results exist for situations where there is a homology of over 50 percent with the target protein. Thus when the level of homology is this high, a direct and general relationship does exist between sequence and structure. Moreover, there are some protein classes for which the relationship between structure and function is common and yet which show no significant overall homology. This raises the intriguing question of whether some other algorithm, unrelated to sequence homology, might exist for uncovering such relationships. Another approach has been to analyse turns in known structures and seek regularities in the sequences within particular structural classes.

Another example of the need for simultaneous access to structure and to sequence databases is provided by research on carbohydrates. The researcher interested in a particular glycoconjugate may investigate the carbohydrate database. He might have the carbohydrate side chains of the glycoprotein and may also know the exact amino acids to which the side chains are attached. But to study the interactions between the protein and carbohydrate chains he must exit the carbohydrate database to search another for the amino acid sequence and yet another for the three-dimensional structure.

Until recently databases had not been organised in a way convenient for combined structure and sequence use. For convenient access, integration of primary data (sequences with physicochemical properties of their residues), secondary data (the assignment of residues to helices and sheets), and tertiary data (interaction between the elements of secondary data and of the regions, called loops, between these items) would be necessary (Lesk, 1989).

8.5.1. The Canadian scientific numeric database service (CAN/SND)

One approach to this problem is the Canadian scientific numeric database service (Wood et. al., 1989). CAN/SND allows the user to gain access to a large variety of international online databases. It is a network which is served by a mainframe on which runs a continuous file-server which is a virtual machine which awakens from its 'sleep' mode in order to service user requests. Together with a wide variety of databases, there are many search and analysis programs. Wood et. al., (1990) provide an interesting application example with the user's queries weaving in and out of databases and modelling software.

8.5.2. The ISIS Integrated Sequence and Structure Database

Another example of an integrated gateway software system allows access to some of the main sequence databases (NBRF-PIR, SWISSPROT, GENBANK, NBRF NEW, NEWAT85, JIPID), and to the Brookhaven structure database. It is called Integrated Sequences Integrated Structures (ISIS) and has been developed by the Protein Engineering Club Database Group (Askigg et al., 1988). Figure 8.2 shows the various components of the system.

The sequence database OWL integrates the publicly available databases and also uses a translated version of GenBank nucleic acid data.

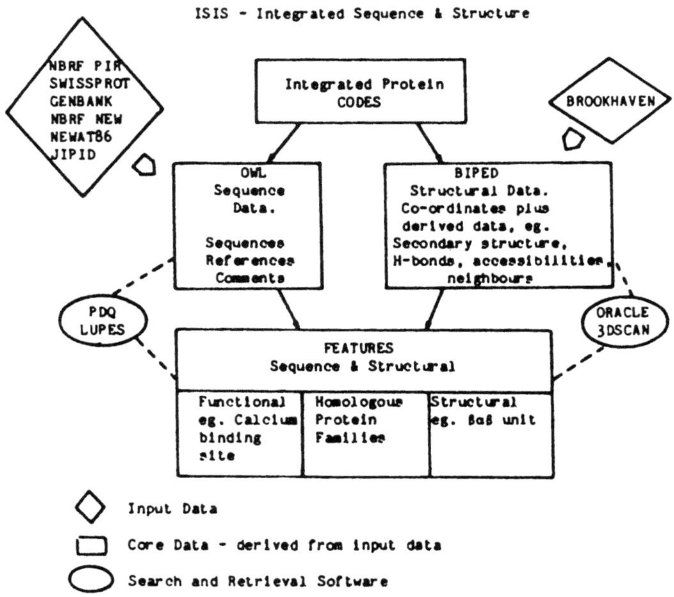

Source data: **NBRF PIR and NEW 16·0:** National Biomedical Research Foundation, Georgetown University Medical Center, 3900 Reservoir Road, N.W., Washington, DC 20007, USA[4]. **SWISS-PROT 6·0:** A. Bairoch, Departement de Biochemie Medicale, Centre Medicale Universitaire, 1211 Geneva 4, Switzerland. **GenBank 54:** Los Alamos National Laboratory, Los Alamos, New Mexico 87545, USA (translations used in part programs of J.W. Fickett, 1986)[5]. **NEWAT 86:** R.F. Doolittle, Department of Chemistry, University of California San Diego, La Jolla, California 92093, USA. **JIPID:** Japanese International Protein Information Database, Science University of Tokyo, Japan. **BROOKHAVEN October 1987:** Brookhaven National Laboratory, Upton, New York 11973, USA.

Figure 8.2. The ISIS integrated protein sequence and structure gateway. (Source: Askigg et al., 1988).
Reprinted by permission from Nature Vol. 335, pp. 745-748, Copyright (c) 1988, MacMillan Magazines Ltd.

PDQ is an established search and retrieval package. LUPES uses pattern discriminators for knowledge-based database exploration. BIPED is a relational database of 3D structures based on the Brookhaven collection. Additional information derived from the 3D coordinates, such as secondary structures, torsion angles, solvent possibilities, hydrogen bonds, salt bridges, disulphide bridges, and nearest neighbours are calculated also. ORACLE is a standard commercial relational database with simple query language (SQL). 3DSCAN is an adaptation of this query language for 3D structure searching.

The usefulness of ISIS can be seen by considering protein motifs. Compared with the number of known chemical compounds (nine or ten million) and the number of proteins (14,000), there are only 400 proteins for which 3D structures are known. Despite this, it is clear that proteins fall into clearly identifiable families, constructed from a limited number of motifs. The restriction to L-amino acids, and also the geometry of peptides, limits the conformational space available to polypeptide chains. As a result, the alpha helix, the beta strand, and composite secondary structure patterns or motifs (for example the beta-alpha-beta unit) dominate protein structures. These motifs are useful in searching for similarities between proteins. Often the similarities which do exist are surprising and bring about great leaps in insight into protein evolution and function. These motifs can be arranged hierarchically (Thornton and Gardner, 1989).

The question of how these motifs can be recognised and combined is therefore an important one, and one which the ISIS system handles with some ease. Because motifs can be built up from sequences and from structures it is necessary to be able to combine these two sources in one system: hence the usefulness of ISIS. Three types of motif have been found useful

(1) Sequence motifs are derived from homologous sequences, for which it may be possible to get structural information. Homologous sequences are discovered by aligning two or more sequences and finding those sequences with most similarities to a target sequence or to a consensus sequence. This process is becoming increasingly automated.

(2) Structure-related sequence motifs are sequence patterns that have been found to co-occur along with a particular structure. For

example many identical pentapeptides adopt different secondary structures, so that the local structure depends upon the context. Despite this some are independent of context.

(3) Structural motifs include the alpha helix, the beta strand, and composite patterns such as the beta-beta hairpin, the beta-alpha barrel (found in 14 proteins so far), and so on.

The database query language 3DSCAN can be used to search for these recurring motifs. Occurrences of motifs can be identified from more than one table at the same time. Also for a given sequence motif, the corresponding structural information can be extracted. For example it is possible to ask for the observed secondary structure for all examples of the sequence Gly-Gly-X-Leu (Thornton and Gardner 1989 found that 12 out of their 13 examples had a X-X-beta-beta conformation). Another example which Thornton and Gardner offer is of finding all examples of a disulphide bridge between two alpha-helices. These were found to be rare: there were only 13 examples out of 300 structures investigated. They also show how ISIS and 3DSCAN can be used to extract much more complicated information relating sequence and structure. The ISIS system can be used to display and superpose examples of "hits" automatically (Thornton and Gardner, 1989).

8.6. REFERENCES

Abola E.E., Bernstein F.C., and Koetzle T.F., 1988, The Protein Data Bank, 69-81 in Lesk A.M., (Ed.), Computational molecular biology: sources and methods for sequence analysis, Oxford University Press, Oxford.

Allen F., Bellard S., Brice M.D., Cartwright B.A., Doubleday A., Higgs H., Hummelink T., Hummelink-Peters B.G., Kennard O., Motherwell W.D.S.,

Rodgers J.R., and Watson D.G., 1979, "The Cambridge Crystallographic Data Centre: computer-based search, retrieval, analysis and display of information," Acta Cryst., B35, 2331-2339.

Allen F.H., Bergerhoff G., and Stevens R., (Eds.) 1987, Crystallographic databases, International Union of Crystallography, Chester, U.K.

Askigg D., Bleasby A.J., Dix N.I.M., Findlay J.B.C., North A.C.I., Parry-Smith D., Wootton J.C.,Blundell T.L., Gardner S.P., Hayes F., Haslam S., Sternberg M.J.E., Thornton J.M., Tickle I.J., 1988, " A protein sequence / structure database", Nature, 335, 20 October, 745-748.

Bernstein F.C., Koetzle T.F., Williams G.J.B., Meyer E.F., Brice M.D., Rodgers J.R., Kennard O., Shimanouchi T., and Tasumi M., 1977, "The Protein Data Bank: a computer-based archival file for macromolecular structures", J. Molec. Biol., 122, 535-542.

Brint A.T., Davies H.M., Mitchell E.M., and Willett P., 1989, "Rapid geometric searching in protein structures", J.Mol.Graphics, 7, 48-53.

C.B.M.I.U.C., 1989, "Policy on publication and the deposition of data from crystallographic studies of biological macromolecules. The Commission on Biological Macromolecules of the International Union of Crystallography Report, International Union of Crystallography", Acta Cryst., B45, 518-519.

Doubet S., Bock K., Smith D., Darvill A., and Albersheim P., 1989, "The comple carbohydrate structure database", Trends in Biochemical Sciences, 14, December ,475-477.

Jansson P.-E., Kenne L., and Widmalm G., 1989, "Computer-assisted structural analysis of polysaccarides with an extended version of CASPER using 1H and 13C-n.m.r. data", Carbohydrate Research, 188, 169-191.

Koetzle T.F., Abola E.E., Bernstein F.C., Bryant S.H., and Weng J., 1989, Collection and standardisation of crystal structure data by the Protein Data Bank 77-81 in Colewell R.R., Biomolecular data: a resource in transition, Oxford University Press, Oxford.

Lesk A.M., 1989, Integrated access to sequence and structural data: principles of design of comprehensive databases for molecular biology, 189-198 in Colewell R.R., Biomolecular data: a resource in transition, Oxford University Press, Oxford.

Mitchell M., Artymiuk P.J., Rice D.W., and Willett P., 1990, "Use of techniques derived from graph theory to compare secondary motifs in proteins", J. Mol. Biol., 212, 151-166.

Protein Data Bank, 1985, Atomic coordinates and bibliographic entry format description, Chemistry Department, Brookhaven National Laboratory, Upton, New York 11973, U.S.A.

Thornton J.M., and Gardner S.P., 1989, "Protein motifs and database searching", Trends. Biochem.Sci. 14, 300-304.

U.S.D.O.E., 1986, Summary report of a workshop on a carbohydrate structure database held at Auburn, New York, August 7-9, U.S. Department of Energy, Division of Biological Energy Research, Washington D.C., DOE/ER-0310.

Wlodawer A., Savage H., and Dodson G., 1989, "Structure of insulin: results of joint neutron and X-ray refinement", Acta Cryst., B45,99-107.

Wood G.H., Rodgers J.R., and Gough S.R., 1989, "Canadian scientific numeric database service", J. Chem. Inf. Comput. Sci., 29, 118-123.

Chapter 9:

Sequence Searching

9.1. INTRODUCTION

The standard situation is where the researcher has a DNA sequence which may or may not code for a known gene product. The sequence is read into a frame, called an open reading frame or 'orf' consisting of a long run of amino acid codons with no terminator codon. The problem is usually to find if this unidentified reading frame or 'urf' really codes for a protein, and to find if this protein is known to exist in nature. Two approaches to this problem exist. Firstly one can do computer searching. Or secondly one can use antibodies raised against synthetic peptides "patterned on the sequence of the expected gene product" (Doolittle, 1986). This section will introduce the first method.

According to Doolittle (1986) the least satisfying result is an exact match with a previously reported sequence, indicating that one has not found anything new. The most satisfying result is where one has found a sequence which has similarities with a scientifically interesting protein such as an oncogene. Less satisfying is a similarity with a less interesting protein such as ribonuclease. The most likely finding is that one's protein does not resemble anything in the databank. This would mean that although one had something unique, one would have to do much more to ascertain whether or not it was really a protein.

The relationship and similarity between proteins is partly determined by how proteins have evolved. A single base can be replaced in a mutation by a substituted subsequence. Also there is a constant process of deletions and insertions. The sequences of two related proteins will thus have their alignments interrupted. This is the 'gap problem'. Figure 9.1 shows how intelligent use of gaps can increase the similarity of two sequences. The sequences are the alpha and beta chains of human haemoglobin. The upper panel has gaps inserted (arrowed) and a higher similarity between the two sequences.

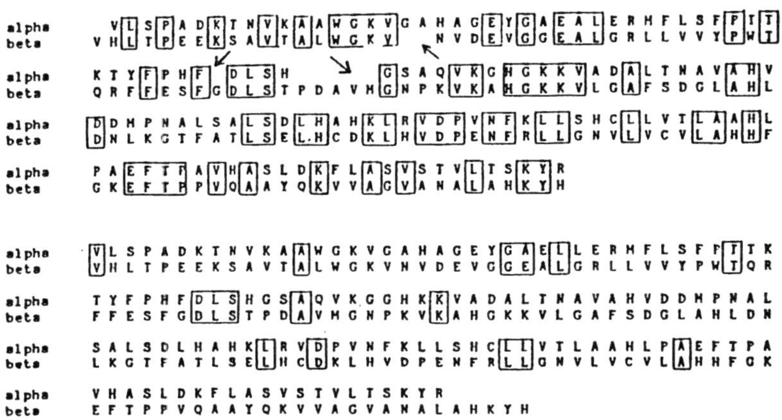

Figure 9.1. Alignment of alpha and beta chains in human haemoglobin. (Source : Doolittle 1986 p.8).
Reprinted with permission from Doolittle R.F., "Of ORFs and URFs: a primer on how to analyse derived amino acid sequences", Copyright (c) 1986, University Science Books.

Also there can be large scale segment rearrangements. For example, the low density lipoprotein receptor has over a large part of its amino-terminal region a segment repeat which is the same as that found in the complement protein C9, but which then goes on to a subsequence which is very similar to the epidermal growth factor precursor.

One reason for the importance of sequence searching for protein similarities is the need to discover the 3D structure of a protein when this is unknown. However, sequence similarity also gives other information. Besides guessing the 3D structure one can also guess protein function. For example, a sequence comparison showed that the milk protein beta-lactoglobin was similar to human serum retinol binding protein. This suggested that a possible function of beta-lactoglobin might be to facilitate the uptake of retinol in young animals (Pervaiz and Brew 1985).

The simplest search algorithm takes a subsequence from each sequence, say a 25 residue subsequence in each case. It compares them, looking for some predetermined level of similarity between them. It then keeps the 25 residues in the unknown protein one is interested in the same, and moves one residue along in the databank sequence. It is now comparing residues 1 to 25 in the unknown sequence with residues 2 to 26 in the databank sequence, and so on. This approach deals with the problems of gaps and rearranged sequences.

Often the desired level of similarity is not predetermined but is derived from the results of the comparisons in a dynamic way. The algorithm counts the number of identical residue pairs, and also weights by the degree of positional alteration it has had to carry out. Because this type of analysis is time consuming usually some initial results are printed.

One specific example of an initial results algorithm is that by Wilbur and Lipman (1983). This is based upon seeking k-in-a-row matches between the two sequences, where k is usually an integer between 1 and 4. Once these have been identified, windows around these regions are compared with an alignment scheme which scores on a similarity scale in such a way that necessary degree of realignment is inversely related to similarity score. When two sequences are compared they are rearranged in a random way a large number of times, these rearranged sequences are compared and the results used to form a distribution. This distribution and its statistics are used to find if the true comparison is significantly different from the mean of this distribution.

A similar idea is used in the FASTp program of Lipman and Pearson (1985). Here the value of k is usually 2, on the assumption that any significant match includes at least 2 matchings. All other segments are eliminated, thus speeding up the search, though reducing sensitivity also. A more general description of this program together with other programs is given by Pearson (1990). FASTp and FASTn are both examples of rapid sequence search programs, and are available on the BIONET network along with programs to align DNA sequences rapidly (ALIGN and PC-ALIGN), programs to minimise free energies in RNA structures (BIOFOLD and PC-FOLD), a suite of sequence analysis programs called DM, a program for sizing DNA fragments (GEL), a collection of programs for DNA and protein sequence comparisons (IDEAS), a program to display RNA structures graphically (MOLECULE), a program which gives the best alignment of three protein sequences (PROT3), a multifunctional program for DNA sequence analysis (SEQAIDII), a program to align many sequences simultaneously (XMULTAN), and a program for calculating hydropathic profiles of proteins (XPROF) (Brutlag and Kristofferson, 1989, p.289).

Another approach when one is seeking unusual patterns in a nucleic acid or regions of strong similarity shared by two or more sequences is to assign scores to sequences or sets of residues determined by such things as charge, volume, hydrophobicity or secondary structure potential. Statistical probability theory is then used to assess areas of high scores to see if unusual patterns are different from those occurring by chance (Karlin and Altschul, 1990).

It is also possible to introduce some information about evolutionary relatedness when this is available. For example an algorithm by Argos (1987) supplements the standard similarity scores with an optimised linear combination of the properties of the amino acids, chosen to give the best alignments of distantly related sequences for which information derived from crystal structures is available.

A group of homologous sequences can be arranged in a family tree which shows their evolutionary relatedness. The algorithms that calculate these trees use as their measure of distance along tree branches the number of nucleotide replacements or else the number of non-identical residues in the sequences. Often a function of these two measures is used. It is the number of accepted point mutations per hundred residues (PAM).

The difference between sequence similarity and sequence homology is that homology is evidence of evolutionary relatedness. However, it is never possible to prove. Nevertheless, a good indicator is to say that sequences which have a match over more than 25 percent of their length are homologous, while those with less than 15 percent are not. Between 15 and 25 percent identity leaves the question open for more investigation. It is important to remember that two randomly chosen (and unrelated) sequences may have a 20 percent identity using one of the standard computer alignment programs. The reason is that the program puts in gaps, and in this way achieves a somewhat artificially high score. Any sequence position can be subject to reverse changes or back-mutations and to multiple hits. For this reason the relationship between evolutionary distance and sequence dissimilarity is characterised by a negative exponential distribution. For this reason, two sequences which are 50 percent different have had 80 hits (or surviving point mutations) per 100 residues. A protein can take 360 such mutations per 100 residues before it becomes unrecognisable.

However, these figures relating to significance are only a guide and there are some well-known examples of the need to consider more than the similarity level. The following come from Doolittle (1981).

(1) Sperm whale myoglobin and lupin leghaemoglobin. This pair showed only 15 percent similarity yet both contain a haem group, have secondary and tertiary structure similarities, and bind oxygen. They are homologues.

(2) The N-terminal and C-terminal halves of bovine liver rhodanese. There is only 11 percent similarity. Yet they have remarkably similar 3D structures. Boswell and Lesk (1988) suggest that this protein arose by gene duplication followed by divergence.

(3) Chymotrypsin and subtilisin. There is only a 12 percent similarity. Yet these molecules have a common catalytic mechanism involving a serine at an active site. They are not similar and not homologous, and provide an example of convergent evolution.

Sometimes it is useful to search for similarities even between short sequences. Sometimes something of real importance is discovered by this means. Figure 9.2 shows the amount of material from purification of melanocyte tumour cell antigen which Brown et al, (1982) had for comparison purposes. A search of the data bank found only one candidate which was transferrin. Brown et al, promptly found that the tumour bound iron with the same strength as did transferrin. It is now known that the protein transferrin is an important factor in making the tumour melanocyte immortal (Doolittle, 1986).

```
----------------------------------------
Tumour antigen, p97     Human transferrin
                        V
                        P
               G        D
               M        K
               E        K
               V -----  V
               R -----  R
               W -----  W
               C -----  C
               A -----  A
               T        V
               S -----  S
               D        E
               ?        H
               E -----  E
----------------------------------------
```

Figure 9.2. Search of short sequences carried out by Brown et al, (1982). Reprinted by permission from Nature Vol. 296, pp. 171-173, Copyright (c) 1982, MacMillan Magazines Ltd.

9.2. FIRST EXAMPLE OF A SEQUENCE SEARCH

Doolittle (1986, Chapter 1) gives an example of a sequence search carried out on his own search program called NEWAT. He had just determined the DNA sequence of something cloned from a Euglena cDNA library by probing with an oligonucleotide based on human randomase, with which he found an open reading frame (a subsequence not including a terminator codon) of 198 amino acids. His example is useful for a beginner because it goes down to the detailed level of suggesting what needs to be typed in (filename, end of file marker, when and how to enter data, how to correct mistakes, etc). What grouping or class of data to search in the database is determined by the search software. The NEWAT program uses a sixfold classification system (prokaryotes but not Escherichia coli, E. coli, virus, eukaryotes but not vertebrate animals, vertebrates but not humans, humans), while for example GenBank uses twelve classes.

In Doolittle's example, the protein is compared with a randomly created sequence. Only one sequence matches well, but this is another randomly-created sequence we shall call sequence R (for random). One protein has a match on 5 out of 484 segments but this is not enough to warrant further interest. The sequence R had 217 residues so a perfect alignment is not to be expected. In fact it is best to trim the sequences down to be more or less the same length. The next step was to find which segments of the sequence R matched with his unknown protein. This was achieved using the Needleman-Wunch algorithm. This algorithm makes allowances for penalties incurred in a particular alignment by gaps being created in order to maximise the hit rate.

One of the things stressed by Doolittle is the need to be prepared for surprises : to make his point he includes a list of 16 startling matches which have been discovered since 1980 (Doolittle 1986 p.33).

Another important point emphasised by him is the need to put results into a context : in terms both of biological function and of evolutionary relatedness. When a sequence is significantly related to another one appearing in the database, then the biological significance of this similarity must be considered. The sequences could be the 'same' protein from different organisms : for example cytochrome c for humans and for fish have an 80% similarity. Or one sequence might represent a relative that has descended after gene duplication. An important question concerns how long ago the divergence began. Figure 9.3 gives some information about the differing rates at which proteins have changed.

Different proteins change at different rates during the course of their evolution. Also different parts of the same protein change at different rates. In general, the exterior parts of a protein change faster than the interior parts, where binding sites and catalytic units are

	Rate of change	Theoretical Lookback time
Pseudogenes	400	45 million years
Fibrinopeptides	90	200 million years
Lactalbumins	27	670 million years
Lysozymes	24	850 million years
Ribonucleases	21	850 million years
Haemoglobins	12	1.5 billion years
Acid proteases	8	2.3 billion years
Cytochrome c	4	5 billion years
Phosphoglycer-aldehyde dehydrogenase	2	9 billion years
Glutamate dehydrogenase	1	18 billion years

Rate of change measured in actual replacements (ARs) per 100 residues.
Useful lookback time - 360 ARs/100 residues.

Figure 9.3. Different rates of change of different proteins. (Source: Doolittle, 1986, p.37).
Reprinted with permission from Doolittle R.F., "Of ORFs and URFs: a primer on how to analyse derived amino acid sequences", Copyright (c) 1986, University Science Books.

often conserved. In proteins with disulphide units, the cysteine residues are among the slowest to change. If a protein is relatively slow to change, like cytochrome c (see Figure 9.3), then it is expected that it will occur in many different organisms.

9.3. SECOND EXAMPLE OF A SEQUENCE SEARCH

This concerns the relationship between the protein encoded in the Arom locus of Aspergillus nidulans and the corresponding proteins of E. coli. The Arom locus of A.nidulans encodes a large, multifunctional polypeptide that catalyses five consecutive steps in the synthesis of aromatic amino acids. In E. coli these steps are catalysed by five enzymes. These enzymes have genes which are not linked. Arom locus sequence has an open reading frame of 4812 base pairs. Although with such a long sequence it is impossible to analyse which portion is responsible for which catalytic activity, it was known that for *E. coli*, AroB codes for 3-dehydroquinate synthase, AroD for 3-dehydroquinase, AroL for shikimate kinase, and AroD for 3-enolpyruvylshikimic acid 5-phosphate (EPSP) synthase. Using sequence searching Hawkins (1987) determined the physical order within the gene of the regions coding for the different enzymatic activities.

Initially, the *E. coli* sequence was compared with the *Arom* sequence. All the DNA sequences were translated to amino acid sequences. Clear homologies were discovered between AroB, AroA, and AroL and various regions of the *Arom* product. Even in the AroD sequence, there was a small (15 percent) but definite homology.

9.4. PROTEIN STRUCTURE PREDICTION

If an unknown sequence resembles a protein which has already been characterised, much can be learnt about it. The first step in this direction was taken in 1967 when a model of lactalbumin was made using the X-ray coordinates for the backbone of hen egg-white lysozyme. Lactalbumin is a protein found in milk and has common ancestry with lysozyme with which it is 35 percent identical.

A homology search in the Protein Data Bank will probably be the first step in such research today. Only a few hundred proteins are completely characterised. The question is, how likely are such a small number to be similar to an unknown sequence in which one happens to be interested? There is good reason to be optimistic about the answer to this question. Firstly there is a rapid rate at which new 3D protein structures are being added to the database. New technical developments

in recombinant DNA studies have led to much larger numbers of purified proteins becoming available for analysis. Secondly, the number of protein types is finite.

One of the central research problems with proteins is the prediction of their 3D structure. The way to proceed is to identify an unknown, hopefully new, protein, then to search a database for homologous proteins. Then these homologues will hopefully have 3D structures which are similar to that of the unknown protein, such that any differences can be deduced by the necessary 3D structure characteristics of the amino acids in the unknown protein. If a protein is identified in the database to be homologous to the unknown sequence in which we are interested, then the way to proceed is by carrying out modelling on the unknown sequence using the 3D coordinates and other information about the known protein derived from the database.This modelling process will not be dealt with here but because of its central importance in protein structure prediction and its reliance on database information, it is dealt with in a separate chapter.

Even where there is no 3D-characterised protein in the database which is homologous with one's unknown sequence, there are several ways of proceeding.

(1) There are a lot of proteins which have repeating sequences: examples include clostridial ferredoxin which has two 26-residue repeats, and E. coli beta-galactosidase, which includes two massive 398-residue repeats. A long sequence with no repeats can be assumed to be either very ancient or changing rapidly (Doolittle, 1986, p.50).

(2) Some patterns indicate particular types of binding properties or active sites. These can be used to create a special subsequence

which is compared with the unknown sequence. These 'consensus sequences' are an attempt to represent by means of ambiguities, variable length gaps, and mixed sequences what is meant by a particular type of sequence.

(3) Although proteins from all organisms have an unexpectedly similar distribution of amino acids, there are some differences that can provide clues. For example, prokaryotes have significantly less cysteine than higher eukaryotes, and some mitochondrial proteins have noticeable codon assignments. For this reason, clues can sometimes be gained by comparing the unknown protein with the global average distribution of amino acids.

(4) Usually sequence analysis starts at the amino-terminal end of the sequence. As each overlapping segment is analysed, its various properties are plotted at the segment midpoint. One of the most useful properties to study is hydropathy, which includes hydrophilicity (an affinity for water) and hydrophobicity (a repulsion from water). Proteins fold in water, and a stable state is reached when the maximum number of hydophilic groups are in contact with water and the maximum number of other sidechains are buried and not on the surface. Also a long (20 or more residues) segment which is highly hydrophobic and which is free of positively-charged residues has membrane-spanning properties. Proteins usually contain a signal sequence which looks similar to a membrane-spanner. This segment, called a 'loader peptide', is proteolytically removed during or just after export of the protein.

(5) Also it is possible to use the fact that there are classes of proteins which have retained their structure and function even

where their sequences show no homology. There are crucial positions on the sequence where constraints exist for the structure to be as it is. Each position may be constrained by a unique amino acid, or be restricted to a class of residues. This pattern of constraint forms a template which can be used in searching the database. An example is the pattern of constraints discovered by Wierenga and Hol (1983). This pattern was associated with the binding of a nucleotide to a beta-alpha-beta unit in a typical nucleotide-binding domain. The pattern discovered was

i o * o G * G * * G * * * i * * i..........i i

where G represents an obligatory glycine, i represents a site that must be occupied by a hydrophilic residue, ... represents an unconstrained region of indefinite length, and * represents an unconstrained position. The pattern template was used in the sequence of the p21 protein encoded by the EJ and T24 bladder carcinoma gene.

(1) Another procedure that is available when no homologous protein is discovered in the database is secondary structure prediction. The central idea is to use information about the distribution of amino acids whose 3D structure is known to calculate the probability that a given segment will be an alpha helix, or a beta sheet, or a turn. As mentioned above, unfortunately this distribution is not very wide and hence of limited value. Although this technique is worth trying, it has only a moderate probability of success (Kabsch and Sander, 1983, found a 56% success rate on 62 test proteins), and can only be used with confidence on soluble, globular proteins and not on fibrous proteins. Due to the moderate success rate the results from secondary structure prediction should be used with other information. Although the

prediction of secondary structures can be enhanced (but only up to a success rate of 60%) by means of sequence motifs of high predictive value, (patterns of the form Ala-X-X-Phe-X-Glu) the problem is that these motifs are hard to find and will only become more easy to find when databases increase in size. This was the main finding of a study by Rooman and Wodak (1988) who used a database of 75 proteins for which 3D structures were known. They found that the small database size limited their ability to find such motifs, because these motifs need to appear frequently enough in the database in order to eliminate poorly predictive patterns.

9.5. REFERENCES

Argos P., 1987, "A sensitive procedure to compare amino acid sequences", J. Molec. Biol., 193, 385-396.

Boswell D.R., and Lesk A.M., 1988, Sequence comparison and alignment the measurement and interpretation of sequence similarity, 161-178 in Lesk A.M., Computational Molecular Biology: Sources and methods for sequence analysis, Oxford University Press,Oxford.

Brown J.P., Hewick R.M., Hellstrom I., Hellstrom K.E., Doolittle R.F., and Dreyer W.J., 1982, "Human melanoma-associated antigen p97 is structurally and functionally related to transferrin", Nature, 296, 171-173.

Brutlag D.L., and Kristofferson D., 1989, BIONET: an NIH computer resource for molecular biology, 287-293 in Colwell R.R., Biomolecular data: a resource in transition, Oxford University Press, Oxford.

Doolittle R.F., 1981, "Similar amino acid sequences : chance or common ancestry", Science, 214, 149-159.

Doolittle R.F., 1986, Of URFS and ORFS : a primer on how to analyse derived amino acid sequences, University Science Books, Mill Valley, California.

Kabsch W.,and Sander C., 1983, "How good are predictions of protein secondary structure?", FEBS Lett., 155,179-182.

Karlin S., and Altschul S.F., 1990, "Methods for assessing the statistical significance of molecular sequnce features by using general scoring schemes", Proc. Natl. Acad. Sci. U.S.A., 87, 2264-2268.

Hawkins A.R., 1987, "The complex Arom locus of Aspergillus nidulans; Evidence for multiple gene fusions and convergent evolution", Current Genetics, 11, 491-498.

Lipman D.J., and Pearson W.R., 1985, "Rapid and sensitive protein similarity searches", Science, 227,1435-1441.

Pearson W.R., 1990, Rapid and sensitive comparison with FASTP and FASTA, 63-129 in Doolittle R.F., Molecular evolution: computer analysis of protein and nucleic acid sequences, Methods in Enzymology, 183, Academic Press, New York.

Pervaiz S.,and Brew K., 1985, "Homology of beta-lactoglobin, serum retinol-binding protein, and protein H.C.", Science, 228, 335-337.

Rooman M.J., and Wodak S.J., 1988, "Identification of predictive sequence motifs limited by protein structure data base size", Nature, 335, 1, 45-49.

Wierenga R.K.,and Hol W.J., 1983, "Predicted nucleotide-binding properties of p21 protein and its cancer-related properties", Nature, 302, 842-844.

Wilbur W.J., and Lipman D.J., 1983, "Rapid similarity searches of nucleic acid and protein data banks", Proc. Nat. Acad. Sci. U.S.A., 80, 726-730.

Chapter 10:

Case Study: Specification of Expert System for Protein Structure Prediction

10.1. INTRODUCTION

The problem of predicting three-dimensional structure from protein sequences is at present attracting much research interest. It has remained an unsolved problem ever since Anfinsen *et al.*, (1961) showed that ribonuclease could be denatured and refolded without loss of enzymatic activity, thus implying that amino acid sequences contained enough information to define the three-dimensional structure of a protein in a particular context (Fasman, 1989). New primary structures are found in large numbers by using DNA sequencing technologies, and prompt the desire to guess at their three-dimensional structures in order to understand their biological function.

It has also been suggested that the problem could be helped by the use of expert systems. This chapter will set out some requirements which such an expert system should meet. It will thus concentrate upon the question of "What" such an expert system should provide to the research scientist. It will not give any attention to the problem of programming algorithms, which is a "How" question which is best deferred until the statement of requirements has been clarified.

The chapter is based upon information supplied as the result of a survey of U.K. and U.S. university research scientists. A hundred scientists were sent questionnaires requesting details on current procedures and problems involved in protein structure prediction. From the answers it was possible to derive some ideas on what an expert system could provide in this field. What follows is a summary of the responses elicited by the questionnaire.

The inclusion of this chapter is justified by the fact that any survey of online molecular databases must describe some of the applications areas in which molecular databases are used. The prediction of three-dimensional protein structure is one of the more important of these areas, which also has the educational advantage of offering an example of how artificial intelligence is being gradually applied in molecular data retrieval and analysis.

The prediction of three-dimensional protein structures is one of the fastest changing and most exciting areas of molecular science. Its importance and relevance can be seen from the following introduction to the topic by Blundell et al., 1987,p.347):

> "Technological developments of industrial, clinical and agricultural importance may be achieved in the coming years by imitation of the interactions between macromolecules and ligands that occur naturally in the living cell. For example, the design of drugs, herbicides and pesticides may be improved from knowledge of the interaction of a molecule with an isolated receptor, enzyme or nucleic acid. The specificity, stability or activity of engineered hormones of clinical importance or enzymes of value to the chemical industry may be improved from knowledge of proteins in general. New peptide and protein vaccines may also be designed from information on antibody antigen binding. In the longer term new molecular electronic devices may be constructed using novel molecules - perhaps proteins - that aggregate in a predetermined way (as they do in living cells) and provide a template for molecules that conduct electrons; these will be the new biological microchips".

Besides the prediction of the structure of individual proteins, work is also proceeding on the three-dimensional structure of large complexes. For example even complexes as large as ribosomes, which contain on average 35 proteins and 2 RNA chains in the large subunit, have been investigated (Yonath and Wittmann, 1989).

10.2. PROBLEM DESCRIPTION

Although there are about 14,000 known proteins, three-dimensional structures are known for only about 400 of them. The problem is partly that many proteins are not crystalline and hence cannot be investigated using X-ray diffraction. Also proteins are large: although they are formed from only 20 amino acids, a protein may contain many hundreds of non-repeating permutations of such amino acids, making the problem of sequential matching of two sequences of amino acids enormous.

The three-dimensional structure is essentially described by the occurrence of particular amino acids in the primary structure. One fortunate source of simplicity is that within amino acids there are only certain known places which serve as binding sites for reactions. But unfortunately proteins are almost never inert: they alter in accordance with their environment and their shape changes as reactions occur when different molecules and atoms together with the protein form a complex.

Although there are, alas, imperfect methods of clamping proteins, this tendency to change shape is a source of much difficulty in experiments. It also means that descriptions of shape are often inaccurate or at least reflections of only a temporary reality. Indeed what we have is a problem not only of describing three-dimensional structure but also of capturing time as well. This problem means that searching for proteins which have similar sequences to a target protein is likely to be fraught with difficulties.

One fortunate general principle associated with proteins is the reason for their stability. This arises mainly from the delicate balance between the reduction in the apolar surface area accessible to solvent that occurs when hydrogen bonds form simultaneously with folding and with the associated entropy loss caused by a well-defined structure. This is an important source of predictability and simplification.

To understand the process of protein folding it is necessary to get structural information about folding intermediates on the folding pathway. This information is difficult to get because intermediates are very dynamic and difficult to study at equilibrium. Methods include the use of disulphide reactions to trap folding intermediates in stable form, and the use of hydrogen isotope exchange and rapid mixing to label transient structural intermediates (Roder et al., 1988).

Besides questions of techniques, the theoretical understanding of the folding pathway remains a difficult challenge. For small proteins, the single-domain proteins barnase and barstar (Hartley, 1989) and ribonuclease A (RNase A) (Udgaonkar and Baldwin, 1988) have traditionally served as a model. In ribonuclease A it seems likely that there is an early hydrogen-bonded intermediate and a late native-like intermediate on the folding pathway of the main unfolded species. However, little is known of the structure or stability of these kinetic intermediates (Udgaonkar and Baldwin, 1988).

The problem central to this chapter is, given a particular protein sequence, to determine its three-dimensional structure. Each solution seems to have more individual features than ones which reflect a unifying theory. However, general elements do exist. These general elements require to be teased out, often as fuzzy "perhaps" deductions rather than hard and fast rules. This is an area abounding in intuition and hunches. Despite this, such problems have proved amenable to expert systems which use probabilities in their production rules to reflect uncertainty. One example would be the MYCIN medical expert system, although commentators point to some special limiting features. These features include the fact that its use of probabilities contravenes probability theory - e.g. probabilities can theoretically sum to more than one - and the point that its success is due more to the fact that conditional probability chains of the form if A and if B and if C....... are not very long, thus minimising the importance of accuracy of probabilities.

10.3. THE MAIN FUNCTIONS OF A STRUCTURE PREDICTION EXPERT SYSTEM

10.3.1. Expert System Components

Most practical expert systems have two main components. We shall use the example of MYCIN, the medical expert system, to exemplify these two components.

(1) The first such component is a knowledge-elicitor, which is used (perhaps in a series of question and answer sessions with an expert in the application area) to build up a knowledge base about the application area. In our case the application area is that of protein structure prediction and protein sequences. For example, the medical expert system MYCIN has a component which allows medical practitioners to enter new medical case histories into the knowledge base. From information which has been entered about symptoms, treatment and the incidence of cure, worsening or complication of the original ailment, it is possible for the results to be analysed in such a way that new or altered rules can be added to the program automatically. This is one of the reasons why languages such as LISP, which treat data including programs as lists, are popular, since such data structures are easy to extend. In this way the second component is constructed.

(2) This is the rule-using component which, when given a symptom description, will prompt for more information (of a narrowly-circumscribed kind) in order to arrive at a diagnosis. Often the prompting for more information involves the medical personnel undertaking fresh inspections or laboratory tests, or simply going back to the patient for more information in a verbal form.

The data which are entered into the knowledge-elicitor remain in the knowledge base, and the fact that symptom X is associated with symptom Y for diagnosis Z is used by the computer programmer to create rules of the form:

```
----------------------------------------------------------------
               IF symptom X is present AND
               IF symptom Y is present THEN
                       give diagnosis Z
               ENDIF
----------------------------------------------------------------
```

However, the rules are often much more complicated involving many layers of nesting. This is a more common type of rule:

```
----------------------------------------------------------------
               IF symptom X is present AND
               IF symptom Y is not present AND
               IF patient has attribute A1 AND
               IF patient does not have attribute A2 AND
               IF symptom X appeared at relative time T1 THEN
                       IF there has been change C1 AND
                       IF there has not been change C2 THEN
                               give diagnosis D1
                       ELSE IF............
                               give diagnosis D2
                       ...........
                       ENDIF

               ENDIF
----------------------------------------------------------------
```

10.3.2. Avoidance of a flat expert system

The following is an example of a "flat" expert system:

```
--------------------------------------------------------------------
        IF condition C1 THEN do action A1
        else IF condition C2 THEN do action A2
        else IF condition C3 THEN do action A3
        else IF ..........
        else IF ..........
        ENDIF
--------------------------------------------------------------------
```

where in the whole of the program only one case holds (one of the N conditions is true). This entails sequential search of these conditions until the first true one is found and takes on average N/2 tests of conditions. However if the expert system is more hierarchically structured as follows:

```
--------------------------------------------------------------------
        IF condition C1 THEN
                IF condition  C11 THEN
                        IF condition C111 THEN do action A111
                        ELSE IF condition C112 THEN do action A112
                        ELSE IF condition C113 THEN do action A113
                        ENDIF
                ELSE IF condition C12 THEN
                        IF condition C121 THEN do action A121
                        .......
                ..........
                        .......
                ..........
                ..........
                ENDIF
        ............
        ENDIF
--------------------------------------------------------------------
```

then search through the program for true conditions is more rapid. Since one of the major practical difficulties of expert systems is their slowness in searching through the myriad of special cases and conditions, such a design consideration is not a trivial one.

10.3.3. Well-definedness

It has often been said that expert systems are very good for complicated diagnosis problems such as medical or technical fault-finding problems which are well-defined in terms of there being one set of things present in the case and another set of things which are absent in the case. An expert system which shows that well-definedness yields rewards in terms of expert systems is DENDRAL, (Lindsay et al, 1980) which analyses mass spectrometer data of rock specimens and uses geological information about co-occurrence of different geological materials in order to predict geological facts such as the existence of precious or energy-bearing minerals. We must therefore ask the question "Is the protein structure-prediction problem well-defined?"

At present the answer is that many steps towards structure prediction have not been automated precisely because they are ill-defined. However, it is these "intuitive gaps", which scientists have to bridge with guesswork, that offer the greatest scope for an expert system. This is because if the knowledge-elicitation module is successful, the information which so far exists for the 400 proteins whose three-dimensional structure is known (out of 14,000 known protein sequences) could be entered in, together with other information which allows experienced and/or clever scientists to make guesses.

This suggests that there are several levels of well-definedness operating:

(1) At the most defined level are the 400 protein sequences for which three-dimensional structures (and hence other characteristics) have been discovered.

(2) At a less well-defined level there are many "families" and subgroups where it is known that empirical associations have been

established. For example, there are predictable secondary structures, recurring sequence and binding patterns (Reynolds et al., 1989; Svensson et al., 1990; Clarke et al., 1989), and different ways of defining homologies (using peptide methods, using homologous proteins, using structural class). It is known that different hydrophobicity values are associated with the different amino acids, and that hydrophobicity plots can be used to predict exposed regions of a protein (where turns can be found), or buried regions (where particular peptide structures of helices, sheets and turns are associated with the simultaneous formation of hydrogen bonds) and to predict transmembrane segments from plot peaks. Membrane studies have recently promised great advances, because although a large number of prediction algorithms exist to suggest possible conformations, it was not until 1985 that the three-dimensional structure of a membrane protein complex was determined, and to date only two structures are known (Fasman and Gilbert, 1990). As more information becomes known, it will be possible to study the receptors for a large number of ligands that control many important functions (Hulme, 1990). Also it is known that peptides can be used to raise antibodies which are able to recognize native structure. As another example, Arbarbanel et. al., 1987) have used heuristic pattern matching algorithms to study the Human Growth Hormone (HGH) and Interleukin-2 (IL-2). This approach required the use of algorithms for folding secondary structure regions into packed three-dimensional structures, using energy minimisation procedures, and was quite successful in predicting large parts of the three-dimensional structure. These as well as other recurring themes have been collected on a one-off basis by researchers tracking down a particular three-dimensional structure. However, they represent a generalisation of results from the 400 or so known three-dimensional structures. What has not yet been done is (a) to put these 400 structures together in order to automate the generalisation process, and (b) to attempt to build these findings into a completely integrated rule-system. However, a start is being made. For example, Dolata and Prout (1987) and Wipke and Hahn (1987) use artificial intelligence techniques to construct a rule base of axioms which provide logical information (derived from both human and machine expertise) which is then used to infer candidate structures These candidate structures are then analysed by other (numerical, energy minimisation) procedures. They claim to have dramatically speeded up candidate structure generation.

(3) Where such generalisations cannot be made, there are still some clues to three-dimensional structure. For example, the absolute conservation of the residues of the active site mean that the geometry of the site has been left unchanged throughout evolution, and so the problem of building the molecular model in this region implies that structure can be used at the atomic level for three-dimensional prediction. In such cases the secondary structure of the core supporting the active site residues is usually left unchanged within a domain and within a particular family. But most of the residues will have had some mutation so that the loop lengths away from the active site would have undergone some change. Beyond this, on the edges of the beta-sheet, even the sequencing will be uncertain, since although beta-strands still exist here, there is usually no correspondence with any protein whose three-dimensional structure is known.

10.4. ELICITATION OF KNOWLEDGE FROM PROTEIN SCIENTISTS.

10.4.1. The Need for Integration.

At present, the problem is approached as shown in Figure 10.1.

Some of these tasks are automated, while others are not. For example, the model building activity (for the case where homologous protein has been found and the structure is known) is now completely automated: Sali *et al*., (1990,p.235) give one automation scheme and Robson *et al*., (1987) provide another example with their expert system LUCIFER.

Even if the whole task defined in Figure 10.1 were to be integrated (defined as putting all these tasks into one program) the scientist would still need to break off at particular points and do an experiment. Integration would tell him at what points he should experiment and what type of experiment would be most appropriate. Once the result of the experiment was fed back in, the integrated set of tasks would continue. At the end of the set of tasks, prediction accuracy would be used to

Figure 10.1: The current approach to structure prediction

```
(1)     search for homologous protein in sequence databanks.

(2a)    if found align sequences, do conservation plot;
                    if structure is known then
                         model build using coordinates
                         of homologous protein
(2b)    if not found or if structure not known then
                    do secondary structure prediction,
                         hydrophobicity plot,
                         amphipathic plot;
                    Use pattern matching techniques using
                         sequence and structural templates.
                         Try to identify probable fold.
(3)     assess accuracy of prediction.
(4)     if insufficiently accurate go to step (1)
                    otherwise stop.
```

advise the scientist what to change in the next run round the set of tasks. So the schema shown above could be viewed as a looping program from which abort or success serve as exit conditions.

An advantage of this approach is that the program "remembers" all parameter values at previous stages of execution (and indeed storing values from yesterday's run or from the day before would also be easy to implement, and a convenient facility to have). If the scientist makes a mistake he can go back to a previous stage of analysis.

A more detailed flow diagram based upon this approach is given by Taylor (1987, p.318).

A different approach is taken by Clark et al., (1990a, p.98). They begin by specifying 29 entities each of which has its own data type. For example, "biological source" is of type "substance", and "DNA sequence" is of type "sequence" and "tertiary structure" is of type "xyz coordinates". There are two types of relationships between these entities: minimal preconditions and constraints. A is a minimal precondition for B if, under some circumstances, there is a process that can be used to derive an hypothesis for B from A. Constraints are consistency requirements between entities. A constrains B if the information contained in A limits the range of possible values or conformations of B in some situation, and A is not a minimal or additional precondition for B. Clark et al., (1990a, p.99) give the following example:

> "Thus, disulphide linkage constrains an alignment because corresponding cysteine residues in the disulphide linkage should be aligned and disulphide linkage is not a precondition for alignment".

Using these definitions, they give an example of the kind of logic a system might carry out (Clark et al., 1990a, p.101):

```
"if X is known
        and X is a minimal precondition of Y by Z
        and Y is not known
then it is possible to derive Y from X by Z"
```

Their initial test program PROPS2 proceeds by outputting at various stages a list of possible actions which the user might take. At some stages the list of alternatives is wide but at others it is constrained by new information which the user has gathered. The calculation of this is automated. PROPS2 can respond to certain requests for information, such as "What are the consistent relationships established thus far?", or "How do you check the accuracy of a predicted tertiary structure?", or "How can an inconsistency be resolved?". Some of the alternatives displayed on the screen are executable - they are further parts of the program - while others are not and require the user going away to do experiments and so on. Although they have not yet implemented such enhancements, Clark *et al*. have suggested that higher-level advice could be built into the list of alternatives to show which are recommended. The logic used to arrive at such recommendations might use such concepts as that of logical inclusion:

```
"if goal is X
        and A is a minimal precondition of X by C
        and B is an additional precondition of X by C
        and constraints of A include B
then derive B before A in context of goal X"
(Clark et al.,1990a, p.104).
```

10.4.2. Graphical Interface

Clark and his colleagues have now developed a more advanced program, PAPAIN (Protein Analysis and Prediction using Artificial Intelligence) (Clark 1990b). One of its features is the use of a sophisticated mouse-driven graphics interface, together with graphical-based tools for allowing the user to directly interact with the growing knowledge base. Logic programming (used in such languages as Prolog) can be used for the derivation of topological descriptions of proteins (Rawlings *et al.*, 1985) and for a flexible query language (Rawlings, 1987) as well as for developing a query language for graphics (Seifert and Rawlings, 1988). PAPAIN includes a set of graphical tools which develop and extend this work. There are several graphics tools.

(1) Analysis tool to do data analysis.

(2) Profile tool to display and annotate sequence analysis profiles.

(3) Cartoon tool to interact with protein topology displays which control browsing and querying.

(4) Ribbon tool for showing protein secondary structures of the protein alpha-carbon backbone.

(5) Browser to move around frame-based generalisation hierarchies of protein structure and function. Frames are classes which have characteristics which can be inherited (unless over-ridden) by class members.

Each tool has its own window (see Clark, 1990b, p.4 for an example screen). Each window uses the same information in the knowledge base. In this context, Clark *et al.*, 1990b, p.4) suggest a specification of an integrated graphics interface:

(1) Each window gives its own view of the same knowledge base information.

(2) Each window offers the facility to the user of being able to be updated in such a way that its updates cause the other windows to be updated.

(3) Each window performs the dual role of allowing data to be updated and the knowledge base to be augmented.

10.4.3. Making structure prediction programs intelligent

Scientists carrying out three-dimensional structure prediction have a "target" protein whose sequence is known but whose three-dimensional structure is unknown. The main hope is to find another, similar, protein (i.e. one with a similar sequence) for which the three-dimensional structure has been found; i.e. the homologous protein is one of the 400 for which three-D structure is known. This is the most critical decision point in the process (see Figure 10.1).

This test of homology is critical in the sense that if no similar sequences are found then scientists are driven to choose the far less satisfactory path beginning with secondary structure prediction (see Figure 10.1) and which includes methods with notoriously low prediction rates (e.g. structural class predicts in only 15% of cases). Getting a homologous match in the protein database is then of great importance. Unfortunately searches can take up to 10 minutes on medium sized machines such as a VAX when the search program has been given no further clues besides the amino acid sequence of the target protein. Searches typically try to match all subsequences with all subsequences, but unfortunately it is possible for there to be gaps in one sequence, so that what is needed is to match all permutations of subsequences of one protein to all permutations of subsequences of another, a quite impossibly slow task. What are required are ways of discovering the reasons why experienced scientists guess at subsequence combinations, and why they guess that a gap will exist. At a more general level this kind of information would be crucial for the development of novel, second-generation databases:

> "A generalised framework of conception and relationships abstracted from human thinking could enormously increase the performance of molecular biology software. New categories identified by computer-based methods could be

> introduced to complement the known substructures. For example, linguistic methods have been used for the automatic definition of characteristic subwords and syntactic rules for DNA. Moreover, artificial intelligence methods could be built up using these extended concepts, for automatic analysis of the available structure data and for building up new structural databases. Novel database structures could be the final result, rather than the beginning of this process" (Pongor, 1988).

During the 1990s it is likely that a new spur to the scale of available scientific data will be encountered. This will be the Earth Orbiting System (EOS) developed by the American National Aeronautical and Space Agency (NASA) for transmitting data by satelite. Any such increase in the scale of data transmission will require methods of abbreviating and summarising the data, using artificial intelligence. Yet the task is not yet solvable, and may never be solvable bearing in mind that such pre-processing of the data assumes one knows what the eventual use of the data is going to be. In most situations, one cannot know what the data will be used for. It may therefore be better to leave data in its vast and raw state. Here one would be relying upon the increasing cheapness of computer memory. Also one would be attempting here and there to economise by squeezing data into the available space using data compression methods (particularly important for pictorial data). One would also be depending upon new and more intelligent approaches to search, again relying on the increasing cheapness of hardware to allow one to parallelise the storage and search for data.

One improvement towards greater "intelligence" in database-searching would be the automated and adaptive use of templates. The first use of this approach was by Taylor and Thornton (1983), who used a template for the ADP-binding beta-alpha-beta fold in nucleotide-binding proteins. Here, it had been noticed that these binding properties are associated with a particular conformation, and this association has been used to increase predictiveness. However, there is no reason why such associations could not be carried out by the program (perhaps at times when the machine was little-used) and the best associations used for finding homologies between proteins with known structure and target proteins with unknown structure.

Some general principles do exist. We consider soluble, globular proteins as an example. It is known that single-residue conformations are retained close to conformations of minimal energy. The shape of the helix and sheet surfaces make these structural elements pack together in only a limited number of relationships. Links between secondary structures tend to be right handed, short and un-knotted. Because of this proteins usually fold into secondary structures comprising only a limited number of folding units or patterns (see Figure 10.2). These include alpha-alpha formation of two anti-parallel packed helices; beta-beta formation of two anti-parallel sheet strands; and beta-alpha-beta formations of a helix packed against two adjacent parallel sheet strands. This means that there are five main structural classes of globular proteins: all-alpha; all-beta; those built from beta-alpha-beta units; alpha plus beta, with helix and sheet regions segregated; and coils, in which there is no secondary structure.

As another example, we consider the structure of membrane proteins, which are as yet very little known. These are determined strongly by the apolar (not electrically charged) nature of the interior of the bilayer membrane. Since polar groups are expected at the membrane boundaries to interact with the polar phospho-lipid molecules, the hydrophobic segment probably is bounded by clusters of electrically charged or polar amino acids. The hydrophobic nature of a protein can be used for displaying the sequences which are most likely to be embedded in the non-polar membrane environment. Also in order for a protein to pass through a hydro-

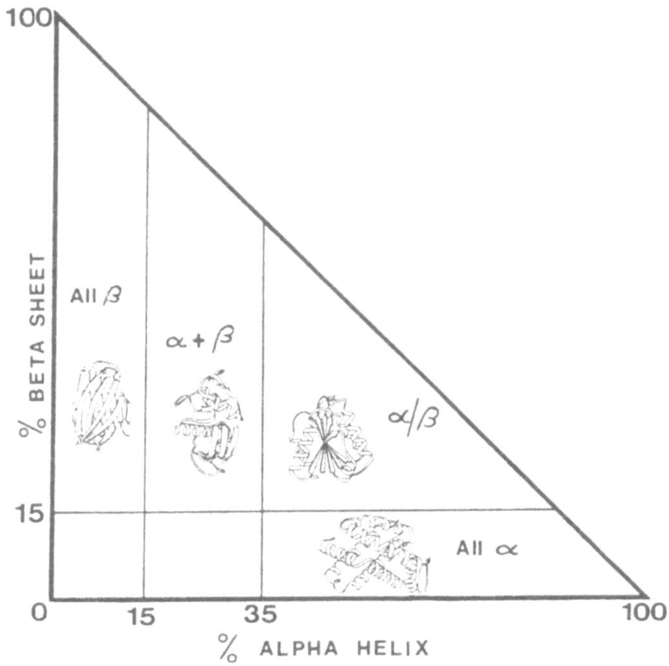

Figure 10.2. Composition graph of secondary structure content (% alpha versus % beta) showing what the four types look like. (Source: Thornton and Taylor, 1989, p.155).
Reprinted with permission from Thornton J.M. and Taylor W.R., "Structure prediction", in Findlay J.B.C. and Geisow M.J., (Eds), "Protein sequencing: a practical approach", Copyright (c) 1989, IRL Press.
By permission of Oxford University Press.

phobic membrane bilayer, an extended sequence of hydrophobic residues is usually required. A molecule in a membrane needs to be spun before it can change from polar to apolar. This illustrates the two ways in which proteins solve the "compatibility problem" Either a protein has long apolar alpha helices, or it has a closed beta structure (a barrel). This allows us to use a two-way classification to characterise secondary structures of this class of proteins.

The above two examples suggest that any program would usefully include the protein characteristics as part of its conditions. The first example shows that single-residue conformations are retained close to conformations of minimal energy. The second example shows that in order for a protein to pass through a hydrophobic membrane bilayer, an extended sequence of hydrophobic residues is usually required.

If a homologous protein is not found then the scientist is forced to consider secondary structure prediction. Examples of the prediction of protein structure without homologous protein (and hence without explicit use of model-building techniques) include the prediction of structures for interferon (Sternberg and Cohen, 1982), interleukin-2 (Cohen et al., 1986), human growth hormone (Cohen and Kuntz, 1987), alpha-unit of tryptophan synthase (Hurle et al., 1987; Crawford et al., 1987), human epidermal growth factor receptor (Fishleigh et al., 1987) and cation transporting ATPases (Taylor and Green, 1989). As suggested above, such an approach is often unsuccessful. However, there are some sources of predictability. Folding propensity of a particular subsequence and the probability of there being a helix, a beta-strand, or a turn, is partly determined by its residues (which have certain "preferred" conformations), and upon the propensity of the resultant surface to pack material. So if we take into account local subsequences, we are interested also in the local nature of the folding forces. However, this degree of generalisation understates the size of the unknown part of the problem awaiting solution.

When the program reaches the end of the loop (i.e. where the accuracy of the prediction is assessed), the program should be able to make some decisions itself if accuracy is not good enough (see Figure 10.1). Here are a few things that it might change in such circumstances:

(1) The definition of homologous should be progressively altered (rather than widened, since one does not want the program's exe-

cution to become slower and slower). Either the scientist is prompted for hunches about subsequence patterns to look for or gaps to allow in sequences, or there must be some way the program acquires experience about progressive refinement of target sequence characteristics over several runs of a program. For example, in the 400 or so identified structures, what aspects of homology were and which ones were not important? It is this kind of assessment of significant clues which the program needs to pick up either by being fed training examples by the scientist, or by generating random ones and noticing the "good" characteristics. One special subclass of this approach is to treat subsequences rather than whole proteins as the information source for the prediction. This involves searching for short segments of protein of unknown secondary and tertiary structure. Often oligopeptides of similar secondary structure exist.

(2) If a homology has been discovered during the database search, the next step is to use the three-dimensional coordinates of the homologous protein to predict those of the target protein. Yet the coordinates of the homologous protein may be influenced by context, as may the coordinates of the target protein. For this reason it is necessary to reconstruct the context (easier to do for the target protein) and to alter the coordinates accordingly. This reconstruction process may be based on probable rather than certain context information.

(3) Several alternative methods of predicting secondary structure (see Figure 10.1) exist. Some are more suitable for some cases than others. Also they make different assumptions. The Chou-Fasman method (Chou and Fasman, 1978) relies on the fact that different amino acids are associated with different structural

types (e.g. sheets or helices). The Garnier-Robson method (Garnier et al., 1978) is based on the observation that the conformational state of a given residue is determined not only by that residue itself, but also by the neighbouring residues. The potential therefore exists slowly, over a period of time, to build up a profile of the types of cases dealt with better by one method rather than by another.

(4) Evidence in generating hypotheses for protein structure prediction comes from very different sources on different occasions (it may be structural class of a protein, membership of a family of aligned sequences, top-down constraints derived from super-secondary structural motifs, or secondary structure prediction techniques). For this reason it is necessary that the extending of the knowledge base be done by bringing in information from many different sources. This means that there must be some formal method (perhaps by means of a mathematical semantic description together with the predicate calculus) for establishing the relationships between information of different types. Clark et al., 1990b) are currently investigating extending their system's reasoning abilities to enable it to deal with many different types of evidence.

(5) One method of secondary structure prediction which is independent and adaptive is the class of pattern-recognition methods. Usually such methods start with the identification of high-density hydrophilic residues. These regions are labelled as "definite turns". Other criteria are used to find "weak turns". Thus regions are found each with a predictable number of residues. Once the turns have been identified the demarcated regions can be searched for structural elements (helices and sheets) and even particular

alpha- and beta- configurations become to a certain extent predictable. These pattern recognition methods are amenable to learning for they can be run any number of times and the performance of their rules can be compared. Highly performing rules can be given higher weights than poorly performing ones. Although often criticised as being opaque and difficult to understand because of the large number of complicated rules embedded inside them, pattern-recognition algorithms do have the ability to alter themselves at each unsuccessful run of the program (i.e. each sweep through the set of tasks described in Figure 10.1). Also, adaptive rules can easily be inserted into such a program. For example, there is sufficient current knowledge about alpha/beta and non-alpha/beta proteins to make guesses about where beta-alpha-beta units might be located. Intelligent algorithms could be created which change depending upon their good or bad performance in any particular case. Another approach might be to create "ideal" beta-alpha-beta units from training examples. New predictions could then be generated which were biased by the ideal. Degree of predictiveness could be used to alter the way the ideal was constructed in an adaptive way. One example in current operation is the construction of a template that contains information about all individual strands in the conserved immunoglobin beta-sheet and which can reliably distinguish between between immunoglobins and other proteins. One problem with pattern-matching algorithms, however, is that the success with which they are used depends greatly upon the size of the database (Rooman and Wodak, 1988) a problem which is likely to disappear as databases become larger.

10.5. REFERENCES

Anfinsen C., Haber E., Sela M., and White F.H., 1961, Proc. Natl. Acad. Sci. U.S.A., 47, 1309-1314.

Arbarbanel R.M., Cohen F.E., and Kuntz I.D., 1987, Assignment of protein three-dimensional structure", Chemical Information Bulletin, 39, 2, 30.

Blundell T.L., Sibanda B.L., Sternberg M.J.E., and Thornton J.M., 1987, "Knowledge-based prediction of protein structures and the design of novel molecules", Nature, 326, 26 March, 347-352.

Chou P.Y., and Fasman G.D., 1978, Ann. Rev. Biochem., 47, 251.

Clark D.A., Burton G.J., and Rawlings C.J., 1990a, "A knowledge-based architecture for protein sequence analysis and structure", J. Mol. Graphics, 8, 94-107.

Clark D.A., Burton G.J., Rawlings C.J., and Archer I., 1990b, "Knowledge-based orchestration of protein sequence analysis and knowledge acquisition for protein structure prediction", Proceedings AAAI Symposium on AI Stanford, Spring, 1990.

Clarke A.R., Atkinson T., and Holbrook J.J., 1989, "From analysis to synthesis: new ligand binding sites on the lactate dehydrogenase framework. Part 1", Trends in Biochemical Sciences, 14, 101-109.

Cohen F.E., Kosen P.A., Kuntz I.D., Epstein L.B., Ciardelli T.L., and Smith K.A., 1986, "Structure-activity studies of interleukin-2", Science, 234, 349-352.

Cohen F.E., and Kuntz I.D., 1987, "Prediction of the three-dimensional structure of human growth hormone, PROTEINS: structure, function and genetics, 2, 162-166.

Crawford I.P., Niermann T., and Kirschner K., 1987, "Prediction of secondary structure by evolutionary comparison: application to the alpha subunit of tryptophan synthase," PROTEINS: structure, function, and genetics, 2, 118-129.

Dolata D.P., and Prout C.K., 1987, "Symbolic reasoning in conformational analysis", Chemical Information Bulletin, 39, 2, 30.

Fasman G.D., 1989, "Protein conformation prediction", Trends in Biochemical Sciences, 14, July, 295-299.

Fasman G.D., and Gilbert W.A., 1990, "The prediction of transmembrane protein sequences and their conformation: an evaluation", Trends in Biochemical Sciences, 15, March, 89-95.

Fishleigh R.V., Robson B., Garnier J., and Finn P.W., 1987, "Studies on rationales for an expert system approach to the interpretation of protein sequence data", F.E.B.S. Letters, 214 (2), 219-225.

Garnier J., Osguthorpe D.J., and Robson B., 1978, J.Mol. Biol., 120,97.

Hartley R.W., 1989, "Barnase and barstar: two small proteins to fold and fit together", Trends in Biochemical Information, 14, 450-454.

Hulme E.C., (Ed.), 1990, Receptor biochemistry: a practical approach, I.R.L. Press, Oxford University Press, Oxford.

Hurle M.R., Matthews C.R., Cohen F.E., Kuntz I.D., Toumadje A, and johnson W.C., 1987, "Prediction of the tertiary structure of the alpha subunit of tryptophan synthase", PROTEINS: structure,function and genetics, 2,210-224.

Lindsay R., Buchanan B.G., Feigenbaum E.A., and Lederberg J., 1980, Applications of artificial intelligence for chemical inference: the DENDRAL project, McGraw-Hill, New York.

Pongor S., 1988,"Novel databases for molecular biology", Nature," 332, 24.

Rawlings C.J., 1987, "Artificial intelligence and protein structure prediction", Proceedings of Biotechnology Information '86, I.R.L. Press, 59-77.

Rawlings C.J., Taylor W.R., Nyakairu J., Fox J., and Sternberg M.J.E., 1985 "Reasoning about protein topology using the logic programming language PROLOG", J. Mol. Graphics, 3 (4), 151-157.

Reynolds C.A., Wade R.C., and Goodford P.J., 1989, "Identifying targets for bioreductive agents: using GRID to predict selective binding regions of proteins", J. Mol. Graphics, 7,103-108.

Robson B., Platt E., Fishleigh R.V., Marsden A., and Millard P., 1987, J. Mol. Graphics, 5, 8-17.

Roder H., Elove G.A., and Englander S.W., 1988, "Structural characterisation of folding intermediates in cytochrome c by H-exchange labelling and proton NMR", Nature, 335, 700-704.

Rooman M.J., and Wodak S.J., 1988, Identification of predictive sequence motifs limited by protein structure data base size", Nature," 335, 1 September, 45-48.

Sali A., Overington J.P., Johnson M.S., and Blundell T.L., 1990, "From comparisons of protein sequences and structures to protein modelling and design", Trends in Biochemical Sciences, 15, 235-240.

Seifert K., and Rawlings C.J., 1988, GRIPE - a graphical interface to a knowledge-based system which reasons about protein topology, in Jones D.M., and Winder R., (Eds), People and Computers IV, British Informatics Society.

Sternberg M.J.E., and Cohen F.E., 1982, "Prediction of the secondary and tertiary structures of interferon from four homologous amino acid sequences", Int. J. Biol. Macromol., 4, 137-144.

Svensson B., Vass I., Cedergren E., and Styring S., 1990, Structure of donor side components in photosystem II predicted by computer modelling", The EMBO Journal, 9(7), 2051-2059.

Taylor W.R., and Thornton J.M., 1983, "Prediction of super-secondary structure in proteins", Nature 301, 10 February, 540-542.

Taylor W.R., 1987, Protein structure prediction , 285-322 in Bishop M.J., and Rawlings C.J., (Eds.), Nucleic acid and sequence analysis : a practical approach, I.R.L Press, Oxford.

Taylor W.R., and Green N.M., 1989 "The predicted secondary structure of the nucleotide-binding sites of six cation-transporting ATPases leads to a probable tertiary fold", Eur. J. Biochem., 179, 241-542.

Thornton J.M., and Taylor W.R., 1989, Structure prediction, 147-190 in Findlay J.B.C., and Geisow M.J., Protein sequencing: a practical approach, Oxford University Press, Oxford.

Udgaonkar J.B., and Baldwin R.L., 1988, "NMR evidence for an early framework intermediate on the folding pathway of ribonuclease A", Nature, 335, 694-699.

Wipke W.T., and Hahn M.A., 1987, "Analogy and intelligence in model building (AIMB)", Chemical Information Bulletin, 39, 2, 30.

Yonath A. and Wittmann H.G., 1989, "Challenging the three-dimensional structure of ribosomes", Trends in Biochemical Sciences 14, 329-335.

Chapter 11:

Appendix

11.1. SOURCE OF INFORMATION

The following list is based upon the LIMB Database Release 2.0, August 1990, (Lawton *et. al.*, 1989). The database includes about 30 fields of data on each database. Here however, there is only the acronym (under 'entry'), the commonly used name ('name.now'), the contact name ('gen.nam') plus address ('gen.add') and phone number('gen.tel').

11.2. ADDRESSES

entry	AANSPII
name.now	Amino Acid and Nucleotide Sequences of Proteins of Immunological Interest
gen.nam	Dr. Elvin Kabat
gen.add	Bldg 8, Room 126 National Institutes of Health Bethesda, MD 20892 USA
gen.tel	(301) 496 0316

entry	AGRICOLA
name.now	AGRICOLA
gen.nam	Reference Branch
gen.add	Room 111
gen.tel	National Agricultural Library 10301 Baltimore Blvd. Beltsville, MD 20705 USA. (301) 344 4479

entry	AIMB
name.now	Database for Researchers in Artificial Intelligence and Molecular Biology

gen.nam	Dr. Lawrence Hunter
gen.add	Bldg 38A, MS-54 National Library of Medicine Bethesda MD 20894 USA
gen.tel	(301) 496 9300

entry	AMINODB
name.now	Amino Acid Database
gen.nam	Dr. Cornelius Froemmel
gen.add	Institut fur Biochemie Hessische Straae 3-4 Berlin DDR-1040 GERMANY
gen.tel	286 2123

entry	BCAD
name.now	BioCommerce Abstracts and Diectory
gen.nam	Mrs. E. Reed
gen.add	Biocommerce Data Ltd Prudential Buildings, 95 High St Slough SL1 1DH UK.
gen.tel	011 44 753 511777

entry	BIOSISCONN
name.now	BIOSIS Connection
gen.nam	BIOSIS Marketing Section
gen.add	BIOSIS 2100 Arch Street Philadelphia, PA 19103-1399 USA.
gen.tel	(215) 587 4800

entry	BIOSISP
name.now	BIOSIS Previews
gen.nam	BIOSIS Marketing Section
gen.add	BIOSIS 2100 Arch Street Philadelphia PA 19103 1399 USA.
gen.tel	(215) 587 4800

entry	BKS
name.now	Biotech Knowledge Sources
gen.nam	Dr. Anita Crafts-Lighty
gen.add	Biocommerce Data Ltd Prudential Buildings 95 High St. Slough SL1 1DH UK.
gen.tel	011 44 753 74201

entry	BMCD
name.now	NIST/CARB Biological Macromolecule Crystallization Database
gen.nam	Joan Sauerwein
gen.add	NIST 221/A323 Gaithersburg, MD 20899 USA.
gen.tel	(301) 975 2208

entry	BMR
name.now	BioMagRes
gen.nam	Beverley Seavey
gen.add	Dept of Biochemistry NMRFAM 420 Henry Mall Madison, Wisconsin 53706 USA
gen.tel	(608) 262 8528

entry	BRD
name.now	Berlin RNA data Bank
gen.nam	Dr. Volker Erdmann
gen.add	Institut fur Biochemistrie-FB Chemie Otto-Hahn-Bau Thielallee 63 D-1000 Berlin 33 GERMANY
gen.tel	011 49 30 838 6002

entry	CARBBANK
name.now	Complex Carbohydrate Structural Database and CarbBank softwar
gen.nam	Dana Smith
gen.add	Complex Carbohydrate Research Center 220 Riverbend Road Athens, GA 30602 USA
gen.tel	404 542 4484

entry	CAS
name.now	CAS ONLINE
gen.nam	CAS Customer Service
gen.add	Chemical Abstracts Review PO Box 3012 Columbus, OH 43210 USA
gen.tel	(800) 848 6538

entry	CASORF
name.now	CAS ONLINE Registry File
gen.nam	CAS Customer Service
gen.add	Chemical Abstracts Service PO Box 3012 Columbus OH 43210 USA
gen.tel	(800) 848 6538

entry	CATGENE
name.now	Domestic Cat Gene Frequencies: a catalogue and bibliography
gen.nam	Tetrahedron Publications
gen.add	Tetrahedron Publications 37 Cheltenham Terrace Newcastle NE65HR UK.
gen.tel	091 265 9228

entry	CCD
name.now	Cambridge Structural Database
gen.nam	Dr. Olga Kennard
gen.add	University Chemical Laboratory Lensfield Road Cambridge CB2 1EW UK
gen.tel	(011 44) 223 336409

entry	CGC
name.now	Caenorhabditis Genetics Center
gen.nam	Dr. Mark Edgley
gen.add	110 Tucker Hall Univ of Missouri Columbia, MO 65211 USA
gen.tel	(314) 882 7384

entry	CSRS
name.now	Compilation of Small RNA Sequences
gen.nam	Dr. Ram Reddy
gen.add	Department of Pharmacology Baylor College of Medicine Hoyston TX 77030 USA
gen.tel	(713) 798 7906

entry	CURRCONTS
name.now	Current Contents(R)
acc.nam	Fulfillment Services
acc.add	Institute for Scientific Information 3501 Market St. Philadelphia PA 19104 USA.
acc.tel	(800) 523 1850 ex 1483 or (215) 386 0100

entry	CUTG
name.now	Codon Usage Tabulation from GENBANK
gen.nam	Dr. Toshimichi Ikemura
gen.add	National Institute of Genetics Mishima, Shizuaka 411 JAPAN
gen.tel	(011 81) 559 75 0771 ext 643

entry	DBIR
name.now	Directory of Biotechnology Information/Resources
gen.nam	Carole Brown
gen.add	ATCC 12301 Parklawn Drive Rockville, MD 20852 USA
gen.tel	(301) 231 5528

entry	DCT
name.now	Drosophila Codon Tables
gen.nam	Dr. Michael Ashburner
gen.add	Department of Genetics Univ of Cambridge Cambridge UK.
gen.tel	(011 44) 233 333 969

entry	DDBJ
name.now	DNA Data Bank of Japan
gen.nam	DNA Data Bank of Japan
gen.add	Laboratory of Genetic Information Analysis National Institute of Genetics Mishima, Shizuoka 411 JAPAN
gen.tel	-

entry	DIALOGGMC
name.now	DIALOG Medical Connection
gen.nam	Marketing Department
gen.add	3460 Hillview Avenue Palo Alto, CA 94304 USA
gen.tel	(800) 334 2564

entry	DRHPL
name.now	Database for the Repository of Human and Mouse Probes and Libraries
gen.nam	Dr. Donna Maglott
gen.add	ATCC 12301 Parklawn Dr. Rockville, MD 20852 USA.
gen.tel	(301) 231 5586

entry	DROSO
name.now	Genetic Variations of Drosophila melanogaster
gen.nam	Dr. Daniel Lindsley
gen.add	Dept of Biology UC.-San Diego La Jolla, CA 92093 USA.
gen.tel	-

entry ECOLI
name.now E.Coli K12 Genome and Protein Database
gen.nam Dr. Akira Tsugita
gen.add Research Institute of Biosciences
Research University of Tokyo
Yamazaki, Noda 278
JAPAN
gen.tel (011 81) 471 23 9777

entry EMBL
name.now The EMBL Nucleotide Sequence Database
gen.nam EMBL Data Library
gen.add European Molecular Biology Laboratory
Postfach 10.2209
Heidelberg 6900
GERMANY
gen.tel (011 49) 6221 387 258

entry EMBOPRO
name.now EMBOPRO
gen.nam Dr. Rainer Stulich
gen.add Dept of Molecular Biophysics
Zentralinstitut fur Molekularbiologie
der Akademie der Wissenschaft der DDR
Robert-Rossle-Strasse 10
Berlin-Buch 1115
GERMANY
gen.tel -

entry ENZYME
name.now The ENZYME Data Bank
gen.nam Dr. Amos Bairoch
gen.add Departement de Biochemie Medicale
CMU
1 Rue Michel Servet
1211 Geneva 4
SWITZERLAND
gen.tel (011 41) 22 61 84 92

entry EPD
name.now Eukaryotic Promoter Database
gen.nam Dr. Philipp Bucher
gen.add Stanford University School of Medicine
Stanford, CA 94305
USA
gen.tel (415) 723 9256

entry GBSOFT
name.now The GenBank Software Clearinghouse
gen.nam Dr. Yuki Abe

gen.add	IntelliGenetics Inc. 700 El Camino Real East Mountain View, CA 94040 USA.
gen.tel	(415) 962 7364
entry	GC
name.now	Gene Communications
gen.nam	Dr. Jorg Schmidtke
gen.add	Institute of Human Genetics Free University Heubnerweg 6 Berlin D-1000 GERMANY
gen.tel	(011 49) (0)30 3203 312
entry	GDB
name.now	Genome Data Base
gen.nam	Dr. Bonnie Maidak
gen.add	1830 E.Monument St., Third Floor Welch Medical Library Baltimore, MD 21205 USA
gen.tel	(301) 955 9656
entry	GDN
name.now	Gene Diagnosis Newsletter
gen.nam	Dr. Jorg Schmidtke
gen.add	Institute of Human Genetics Free University Heubnerweg 6 Berlin D-1000 GERMANY
gen.tel	(011 49) (0)30 3203 312
entry	GENBANK
name.now	-
gen.nam	Dr. Yuki Abe
gen.add	IntelliGenetics Inc. 700 El camino Real East Mountain View, CA 94040 USA
gen.tel	(415) 962 7364
entry	GRIN
name.now	Germoplasm Resources Information Network
gen.nam	Quinn Sinnott
gen.add	USDA-ARS-BA-PSI-GSL Barc West Beltsville, MD 21046 USA
gen.tel	(301) 344 1666

entry	HDB
name.now	Hybridoma Data Bank: A data Bank on Immunoclones
gen.nam	Catherine West
gen.add	Hybridoma Data Bank
	ATCC
	12301 Parklawn Dr.
	Rockville MD 20852
	USA
gen.tel	(301) 231 5528

entry	HGIR
name.now	Human Genome Information Resource
gen.nam	Dr. Jim Fickett
gen.add	T-10, MS K710
	Los Alamos National Laboratory
	Los Alamos, NM 87545
	USA
gen.tel	(505) 665 0479

entry	HGMCR
name.now	NIGMS Human Genetic Mutant Cell Repository
	Catalog of Cell Lines and DNA Samples
gen.nam	Dr. Richard Mulivor
gen.add	Coriell Inst. for Medical Research
	Camden, NJ 08103
	USA
gen.tel	(609) 966 7377

entry	HGML
name.now	The Howard Hughes Medical Institute Human
	Gene Mapping Library
gen.nam	Dr. Iva Cohen
gen.add	Human Gene Mapping Library
	25 Science Park
	New Haven, CT 06511
	USA
gen.tel	(203) 786 5515

entry	HIVSSA
name.now	HIV Sequence and Sequence Analysis Database
gen.nam	Kersti MacInnes
gen.add	T-10,MS K710
	Los Alamos National Laboratory
	Los Alamos, NM 87545
	USA.
gen.tel	(505) 667 7510

entry	ILDIS
name.now	International Legume Database and Information
	Service
gen.nam	Dr. Sue Hollie
gen.add	Biology Dept, Building 44
	ILDIS Coordination Centre

	University of Southampton Southampton SO9 5NH UK
gen.tel	(011 44) 703 581910

entry	IUDSC
name.now	Indiana University Drosophila Stock Center Stock List
gen.nam	Kathy Matthews
gen.add	Dept of Biology University of Indiana Bloomington, IN 47405 USA
gen.tel	(812) 855 5782

entry	JIPIDB
name.now	Biological Database: Asian and Oceania Node of the International Protein Information Database
gen.nam	Dr. Akira Tsugita
gen.add	Research Institute for Biosciences Science University of Tokyo Yamazaki, Noda 278 JAPAN
gen.tel	(011 81) 471 23 9777

entry	JIPIDM
name.now	NMR Database on Biopolymers: Asian and Oceania Node of the International Protein Information Database
gen.nam	-
gen.add	-
gen.tel	-
comment	This database no longer exists. It has been subsumed into JIPIDV

entry	JIPIDN
name.now	Natural Variant Database :Asian and Oceania Node of the International Protein International Database
gen.nam	-
gen.add	-
gen.tel	-
comment	This database no longer exists. It has been subsumed into JIPIDV

entry	JIPIDP
name.now	Physical Property Database
gen.nam	Dr.Akira Tsugita
gen.add	Research Institute for Biosciences Science University of Tokyo Yamazaki, Noda 278 JAPAN
gen.tel	(011 81) 471 23 9777

entry	JIPIDS
name.now	Protein Sequence Database: Asian and Oceania Node of the International Protein Information Database
gen.nam	Dr. Akira Tsugita
gen.add	Research Institute for Biosciences Science University of Tokyo Yamazaki, Noda 278 JAPAN
gen.tel	(01181) 471 23 9777

entry	JIPIDSN
name.now	Nucleic Acid Sequence Database
gen.nam	Dr. Akira Tsugita
gen.add	Research Institute for Biosciences Science University of Tokyo Yamazaki, Noda 278 JAPAN
gen.tel	(011 81) 471 23 9777

entry	JIPIDV
name.now	Variant Database: Asian and Oceania Node of the Internation Protein Information Database
gen.nam	Dr. Akira Tsugita
gen.add	Research Institute for Biosciences Science University of Tokyo Yamazaki, Noda 278 JAPAN
gen.tel	(011 81)471 23 9777

entry	LIMB
name.now	Listing of Molecular Biology Databases
gen.nam	LIMB Database
gen.add	T-10,MS K710 Los Alamos National Laboratory Los Alamos, NM 87545 USA
gen.tel	(505) 667 9455

entry	LIPIDPHASE
name.now	Lipid Phase Database
gen.nam	Prof. Martin Caffrey
gen.add	Dept of Chemistry, 120 W 18th Ave. Ohio State University Columbus OH 43210 USA
gen.tel	(614) 292 8437

entry	LYSIS
name.now	Protelysis Database
gen.nam	Dr. Borivoj Keil
gen.add	Institut Pasteur 28 Rue du Docteur Boux

	Paris, Cedex 15 75724
	FRANCE
gen.tel	1 6907 9535
entry	MBCRR
name.now	MBCRR Protein Family Diagnostic Pattern Database and Search Tool
gen.nam	Dr. Randall Smith
gen.add	LG-1
	MBCRR/Dana-Farber Cancer Institute
	44 Binney St.
	Boston, MA 02115
	USA
gen.tel	(617) 732 37 46
entry	MEDLINE
name.now	MEDLINE and Backfiles
gen.nam	MEDLARS Management Section
gen.add	8600 Rockville Pike
	National Library of Medicine
	Bethesda, MD 20894
	USA
gen.tel	(800) 638 8480
entry	MICIS
name.now	Microbial Culture Information Service
gen.nam	Dr. Geraldine Alliston
gen.add	Laboratory of the Governemnt Chemist
	Cornwall House Watreloo Road
	London SE1 8XY
	UK
gen.tel	(011 44) 1 211 8834
entry	MICROGERM
name.now	Microbial Germplasm Database and Network
gen.nam	Dr. Larry Moore
gen.add	Dept of Botany and Plant Pathology
	Oregon State University
	Corvallis, OR 97331
	USA
gen.tel	-
entry	MINE
name.now	Microbial Information Network Europe
gen.nam	K.Wads
gen.add	CAB International Mycological Institute
	Ferry Lane
	Kew, Surrey TW9 3AF
	UK
gen.tel	44 1940 4086

entry	MIPS
name.now	Martinsreid Institute for Protein Sequence Data
gen.nam	Dr. Hans-Werner Mewes
gen.add	MPI/GEN Max Plank Inst. fur Biochemie Martinsreid 8033 GERMANY
gen.net	mewes@dm0mpb51.bitnet

entry	MOUSE
name.now	List of Mouse DNA Clones and Probes
gen.nam	Dr. Joseph Nadeau
gen.add	The Jackson Laboratory Bar Harbour ME 04609 USA
gen.tel	-

entry	MOUSEMAN
name.now	Linkage and synteny homologies between mouse and man
gen.nam	Dr. Joseph Nadeau
gen.add	The Jackson Laboratory Bar Harbour ME 04609 USA
gen.tel	-

entry	MSDN
name.now	Microbial Strain Data Network
gen.nam	MSDN Secretariat
gen.add	Cambridge University Institute of Biotechnology 307 Huntington Road Cambridge University Cambridge CB3 0JX
gen.tel	(011 44) 223 277 502

entry	NAPRALERT
name.now	NAPRALERT
gen.nam	Dr. Chris Beecher
gen.add	PCRPS/8ss S. Wood St. Univ of Illinois at Chicago-Medical Center Chicago IL 60612 USA
gen.tel	(312) 996 9035

entry	NEWAT
name.now	NEWAT
gen.nam	-
gen.add	-
gen.tel	-
comment	This database no longer exists

entry	OLIGONUC
name.now	Chemically Synthesised Oligonucleotide Database
gen.nam	Dr. Flavio Ramalno-Ortigao
gen.add	Sekton Polymere Universitat Ulm Ulm D-7900 GERMANY
gen.net	ORTIGAO@DULRUU51.bitnet

entry	OMIM
name.now	Mendelian Inheritance in Man:McKusick database
gen.nam	Dr. Bonnie Maidak
gen.add	1830 E.Monument St., Third Floor Welch Medical Library Baltimore MD 21205 USA
gen.tel	(301) 955 6641

entry	PDB
name.now	Protein Data Bank
gen.nam	Mrs Frances Bernstein
gen.add	Chemistry Dept Brookhaven national Laboratory Upton, NY 11973 USA
gen.tel	(516) 282 4382

entry	PIR
name.now	National Biomedical Research Foundation Protein Identification
gen.nam	Dr. Kathryn Sidman
gen.add	National Biomedical Research Foundation 3900 Reservoir Road, NW Washington, DC 20007 USA
gen.tel	(202) 687 2121

entry	PKCDD
name.now	Protein Kinase Catalytic Domain Database
gen.nam	Anne Quinn
gen.add	-
gen.tel	(619) 453 5218
gen.net	QUINN@SALK-SC2.SDSC.EDU

entry	PMD
name.now	Protein Mutant Database
gen.nam	Ken Nishikawa
gen.add	Protein Engineering Research Institute 6-2-3 Furuedai Suita, Osaka 565 JAPAN
gen.tel	-

entry	PPR
name.now	Plasmid Prefix registry
gen.nam	Dr. Esther Lederberg
gen.add	Sherman Fairchild Building - 5402 Stanford University School of Medicine Stanford, CA 94305 USA
gen.tel	(415) 723 1772

entry	PRCTR
name.now	Plasmid Reference Center Transposon Registry
gen.nam	Dr. Esther Lederberg
gen.add	Sherman Fairchild Building - 5402 Stanford University School of Medicine Stanford CA 94305 USA
gen.tel	(415) 723 1772

entry	PRFLITDB
name.now	PRF/LITDB
gen.nam	Dr. Yasuhiko Seto
gen.add	Protein Research Foundation 4-1-2 Ina Minoh-shi, Osaka 562 JAPAN
gen.tel	072 29 2040

entry	PRFSEQDB
name.now	PRF/SEQDB
gen.nam	Dr. Yasuhiko Seto
gen.add	Protein research Foundation 4-1-2 Ina Minoh-shi, Osaka 562 JAPAN
gen.tel	072 29 2040

entry	PROSITE
name.now	PROSITE
gen.nam	Dr. Amos Bairoch
gen.add	Departement de Biochimie Medicale CMU 1 Rue Michel Servet 1211 Geneva 4 SWITZERLAND
gen.tel	(011 41) 22 61 84 92

entry	PSEQIP
name.now	PseqIP
gen.nam	Dr. Isabelle Sauvaget
gen.add	Institut Pasteur
gen.tel	28 Rue du Docteur Roux Paris, Cedex 15 75724 FRANCE

entry	PSS
name.now	Protein Secondary Structure Database
gen.nam	Dr. Akira Tsugita
gen.add	Research Institute for Biosciences Science University of Tokyo Yamazaki, Noda 278 JAPAN
gen.tel	(011 81) 471 23 9777

entry	PTG
name.now	Protein Translation of GENBANK
gen.nam	Dr. Christian Burks
gen.add	T-10,MS K710 Los Alamos National Laboratory Los Alamos, NM 87545
gen.tel	(505) 667 6683

entry	QTDGPD
name.now	Quest 2D Gel Protein Database
gen.nam	Gerald Latter
gen.add	Cold Spring Harbor Laboratory Box 100 Cold Spring Harbor, NY 11724 USA
gen.tel	(516) 367 8356

entry	RED
name.now	Restriction Enzyme database
gen.nam	Dr. Rich Roberts
gen.add	Cold Spring Harbor Laboratory PO Box 100 Cold Spring Harbor, NY 11724 USA
gen.tel	(516) 367 8388

entry	RFLPD
name.now	CEPH Public Database
gen.nam	Rob Cottingham
gen.add	Centre d'Etude du Polymorphisme Humain 27 rue Juliette Dodu Paris 75010 FRANCE
gen.tel	011 33 1 42 49 98 67

entry	SEQANALREF
name.now	Sequence Analysis Literature Reference Data Bank
gen.nam	Dr. Amos Bairoch
gen.add	Departement de Biochimie Medicale CMU 1 rue Michel Servet 1211 Geneva 4 SWITZERLAND
gen.tel	(011 41) 22 61 84 92

entry	SIGPEP
name.now	SIGPEP
gen.nam	Dr. Gunnar von Heijne
gen.add	Theoretical Physics
	Royal Institute of Technology
	Stockholm S-10044
	SWEDEN
gen.tel	(011 46) 8 787 7172

entry	SIGSCAN
name.now	Signal Scan
gen.nam	Dan Prestridge
gen.add	T-10,MS K710
	Los Alamos National Laboratory
	Los Alamos, NM 87545
	USA
gen.tel	(505) 665 2958

entry	SRRSD
name.now	16S Ribosomal RNA Sequence Database
gen.nam	Dr. George Fox
gen.add	Dept of Biochemical Science
	Univ. of Houston
	4800 Calhoun
	Houston, TX 77004
	USA
gen.tel	(713) 749 3980 or (713) 749 2830

entry	SRSRSC
name.now	Small Ribosomal Subunit RNA Sequence Compilation
gen.nam	Dr. Rupert De Wachter
gen.add	Departement Biochemie
	Universiteit Antwerpen (UIA)
	Universiteitsplein 1
	B-2610 Antwerpen
	BELGIUM
gen.tel	(011 32) 3 82022319

entry	SVFORTYMUT
name.now	SV40 Large T Antigen Mutant Database
gen.nam	Dr. James Pipes
gen.add	Dept of Biological Sciences
	Univ. of Pittsburgh
	Pittsburgh PA 15260
	USA
gen.tel	(412) 624 4691

entry	SWISSPROT
name.now	SWISS-PROT Protein Sequence Data Bank
gen.nam	Dr. Amos Bairoch
gen.add	Departement de Biochimie Medicale
	CMU

	1 rue Michel Servet 1211 Geneva 4 SWITZERLAND
gen.tel	(011 41) 22 61 84 92

entry	TFD
name.now	Transcription Factor Database
gen.nam	David Ghosh
gen.add	Building 38A NCBI/NLM/NIH 8600 Rocksville Pike Bethesda MD 20894 USA
gen.net	ghosh@ncbi.nlm.nih.gov

entry	TOXNET
name.now	Toxicology Dta Network
gen.nam	Dr. Philip Wexler
gen.add	8600 Rockville Pike National Library of Medicine Bethesda, MD 20894 USA
gen.tel	(301) 496 6531

entry	TRF
name.now	The Taxonomic Reference File at BIOSIS
gen.nam	Dr. Robert Howey
gen.add	BIOSIS 2100 Arch Street Philadelphia, PA 19103-1399 USA
gen.tel	(215) 587 4917

entry	TRNAC
name.now	tRNA Compilation
gen.nam	Dr. Mathias Sprinzl
gen.add	Dept of Biochemistry University of Bayreuth Universitatsstrasse 30 Bayreuth D-8580 GERMANY
gen.tel	(011 49) 921 552 668

entry	VECTOR
name.now	Cloning Vector Sequence Database
gen.nam	Dr. William Gilbert
gen.add	Room 211 Whitehead Institute Nine Cambridge Center Cambridge, MA 02142 USA
gen.tel	(617) 258 5139

11.3. REFERENCES

Lawton J.R., Martinez F.A., and Burks C., 1989, "Overview of the LIMB Database", Nucleic Acids Research, 17, 5885-5899.

Index

Brookhaven National Laboratory,
Upton, Long Island, NY (Ed.)

Protein Data Bank CD-ROM

1991. CD-ROM and handbook
DM 998,– ISBN 3-540-14101-4
University: DM 498,–
ISBN 3-540-14102-2

No more queuing in the network, no more quibbling with the I&D department!

You are working in the field of biochemistry or molecular biology, you are using an industry-standard PC, you have powerful software at your disposal for the visualization of macromolecular structures, and you have been waiting for a comprehensive database of proteins, enzymes and polynucleic acids, accessible from your PC at your bench and at any time? Here it is: The **Protein Data Bank CD-ROM** comprises all the files of the original Protein Data Bank which has been compiled and so far distributed on tapes (DATAPRTP and DATAPRFI) by the Brookhaven National Laboratory, Upton, NY (USA). This particular issue contains the 554 atomic coordinate entries, the source codes and bibliographic records of the Data Bank release of July 1990.

The CD-ROM has been produced in High Sierra standard and can be read by all brands of CD-ROM drives complying with this standard. Although the file names follow the MS-DOS convention, the file structure can be read and interpreted by any operating system (MS-DOS, UNIX, MacIntosh-OS, etc.), independently of hosts, data lines and with no costs for online searchers.

Do you go in for molecular modelling? if so, we have a software package which will prove particularly useful to you: MOBY. MOBY can display structures with up to 2000 centres (or atoms), and its tools for probing into and modifying these structures help you to perform the decisive steps of your analyses right at your desk and may even take you as far as the question on hand demands.

Springer-Verlag
Berlin
Heidelberg
New York
London
Paris
Tokyo
Hong Kong
Barcelona
Budapest

MIX
Papier aus verantwortungsvollen Quellen
Paper from responsible sources
FSC® C105338

If you have any concerns about our products,
you can contact us on
ProductSafety@springernature.com

In case Publisher is established outside the EU,
the EU authorized representative is:
Springer Nature Customer Service Center GmbH
Europaplatz 3, 69115 Heidelberg, Germany

Printed by Libri Plureos GmbH
in Hamburg, Germany